OVERHAUL

RICHARD FLINT &
SHIRLEY CUSHING FLINT

A Social History of
the Albuquerque
Locomotive
Repair Shops

University of New Mexico Press | Albuquerque

ISBN 978-0-8263-6249-0 (paper)
ISBN 978-0-8263-6250-6 (e-book)

Library of Congress Cataloging-in-Publication data is on file
with the Library of Congress

Cover photograph courtesy of the Library of Congress,
Prints and Photograph Division
Designed by Felicia Cedillos
Composed in Sabon 10.25/14.25

Contents

Illustrations

Maps

Figures

Tables

Introduction

It has been more than sixty-five years since the last overhauled steam locomotive left an erecting bay at the Albuquerque Locomotive Repair Shops and returned to road service on the Santa Fe Railway. After the intervening years, marked by at least three human generations, only a few shopworkers remain who have firsthand memory of the painstaking work that went on at the Shops—and even secondhand memories are fading. Those dimming memories are often of huge neglected and decaying buildings: broken glass littering the ground and graffiti-tagged walls, interspersed with piles of junked metal and other debris.

Largely forgotten are the Shops' heyday, the seventy-five years between 1880 and 1955, when the Atchison, Topeka & Santa Fe (AT&SF) Railway was far and away the largest employer in Albuquerque, paying solid, consistent wages that supported thriving middle-class communities on both sides of the Santa Fe main line to California. By the 1960s and 1970s, though, the barrios of Barelas (to the west) and San José (to the east) were rundown and in the process of abandonment. Urban renewal projects, designed to eradicate the blight but underfunded, left dilapidated buildings and gap-toothed cityscapes.

Fewer and fewer Albuquerqueans remember the blocks of well-kept houses and neat gardens that preceded the rundown barrios of the Shops district. Neither do they remember that steady railroad incomes of hundreds of families over generations established and solidified the middle-class status and expectations of thousands of workers and their families. Although not always as rosy as that sounds, the wages of many shopworkers raised their own and their children's and grandchildren's positions in society to levels their forebears could only dream of.

Among the descendants of AT&SF shopmen and women whom we interviewed for this book, there are the mother of a US congresswoman, a former chief justice of the New Mexico Supreme Court, lawyers, teachers, engineers, successful and well-known businesspeople, popular sports figures, many college graduates, proud members in good standing of the Albuquerque and New Mexico communities, as well as more distant communities. As has been apparent from study of other US industrial-manufacturing cities, steady work and rising incomes from the late nineteenth century until the middle of the twentieth fostered widespread optimism and provided the wherewithal to realize many of the hopes and plans of AT&SF Shop families.

One of our primary objectives in writing *Overhaul* has been to lay out the reciprocal effects that the resident Hispanic and Native American population of the Middle Rio Grande Valley and the Atchison, Topeka & Santa Fe Railway have had on each other, both positive and negative. The extended encounter and accommodation of resident peoples with a major industry has strongly affected both. Albuquerque's uniqueness among New Mexico communities arose and has been reinforced because of the coming and development of the AT&SF Locomotive Repair Shops.

Originally all but excluded from potential benefits of the overnight arrival of industrialization in their midst, Hispanos and Native Americans came to comprise the majority of skilled workers at the Shops by the mid-twentieth century. A very significant result of incorporation of large numbers of Hispanos and Native Americans into the skilled AT&SF workforce was the rise of a substantial and stable middle-class population in the Barelas and San José neighborhoods of Albuquerque. That middle-class population gradually expanded and spread to other areas of the city, the state, and the country. The effect on generations of descendants of AT&SF shopworkers has been transformational, a fact that we detail with many examples.

Like many other American industrial cities, Albuquerque suffered a serious, but not irreparable, blow when the major industry, in this case the Locomotive Repair Shops, closed in the 1970s. Albuquerque though, unlike many other industrially based cities, transitioned within just a few years into a financial and supply hub for oil, gas, and mining activity within the state, which boomed from the 1950s to the 1980s, and is currently experiencing another upsurge. Furthermore, Albuquerque's economic base diversified through burgeoning involvement in atomic and alternative energy research and development; federal land management; support of tribal communities, higher education, and healthcare; manufacturing related to computerization and digitalization of many aspects of American life; and, most recently, as a film production and coordination center.

What started it all was establishment in the 1880s of a major steam locomotive repair facility and the foundation of modern Albuquerque. It is the story of the people, the place, and the work of the Shops that fills the pages of this book. The sources we used to compile that story range from interviews with former shopworkers and their descendants to photographs and site plans; from annual reports of the AT&SF Railway Company to contemporaneous newspaper accounts of daily happenings at the Shops; from meticulous studies by professional historians and social scientists to engineering specifications and manuals; from city directories and census data to architectural analyses; from the records of major strikes to payrolls for machinists and boilermakers; from reports of the US Geological Survey to issues of the *Santa Fe Employees' Magazine*.

The point is that we have incorporated into *Overhaul* and have taken into account a wide range of sources in an effort to provide the most rounded and diverse view of what made up life and work in and around the Shops over its seventy-five-year lifespan. *Overhaul* is a record of many things that were once common knowledge to the residents of Albuquerque.

For decades, the city set its clocks and determined its rhythm by the steam whistle at the Shops. Everyone knew that the Shops were the clanging heart of Albuquerque. They set the city apart from its neighbors. From the beginning, they represented the energy and innovation that were driving modern life. The industrial virtuosity that was demonstrated by the best machinists produced a practical art that was second to none. Steam locomotives—great, black, clockwork machines that comprised intricate assemblages of thousands of matched and mated parts—required the creative skilled labor of integrated teams of journeymen, apprentices, and helpers in a variety of crafts to overhaul and restore them periodically, and sometimes to rebuild them from the ground up. That work and the people who performed it are what we celebrate here.

CHAPTER 1

Albuquerque and Western Steam Railroading in the 1870s

From the time of its formal founding in 1706, Albuquerque was primarily a modest farming community. As Brian Luna Lucero observes, "Residents of the Middle Rio Grande Valley considered Albuquerque as one of a dozen villages along the river, and often not the most important one."[1] For about 173 years, Albuquerque changed almost imperceptibly. A description of the town written by the Franciscan friar Atanasio Domínguez in 1776 would apply broadly to the town at virtually any moment before 1880: "[It] consists of twenty-four houses near the mission. The rest of what is called Albuquerque extends upstream to the north [along the Rio Grande], and all of it is a settlement of ranchos on the meadows of the said river for the distance of a league from the church." In total, Domínguez reported a population of 763.[2] Albuquerque's population was estimated still at only 800, 70 years later.[3]

There was a small segment of Albuquerque's population that engaged in trade at least part time. Manufactured and exotic goods were carried to Albuquerque and the rest of New Mexico by long-distance overland transport. For many decades, such imports were brought by mule train, *carreta* (cart), and wagon from the Mexican cities of Chihuahua, Durango, and Parral, and the mining areas of north-central Mexico, as well as points farther south. But once Mexico gained independence from Spain in 1821, merchant traffic with the United States steadily increased. New Mexican merchants, including some from

Albuquerque, became regular players in the Missouri-Santa Fe-Chihuahua trade. In October 1846, Susan Shelby Magoffin, whose husband was involved in the Chihuahua trade from the United States, briefly visited the Albuquerque store of Rafael Armijo. She recorded this description in her journal: "The building is very spacious, with wide portals in front. Inside is the patio, the store occupying a long room on the street—and the only one that I was in. This is filled with all kinds of little fixings, dry goods, groceries, hard-ware, etc."[4] During the middle decades of the nineteenth century, Rafael, his brother Manuel, and cousin Salvador, all doing business in Albuquerque, were among the most prosperous merchants in the territory. They had business connections throughout New Mexico and across the United States and northern Mexico. And they were joined in long distance wholesale and local retail trade by at least six other merchant members of their family.[5]

As American trader Josiah Gregg made clear in the 1840s, the staple of trade goods carted from the United States to New Mexico was fabrics, especially "cottons, both bleached and brown."[6] Nevertheless, "the economy of New Mexico [including Albuquerque] at mid-century operated mostly at a subsistence level."[7]

That situation began to change significantly with occupation of New Mexico by the US Army in 1846. The stationing of between 700 and 1,000 US troops in New Mexico—soon to be a US territory—and the army's renting of buildings as post quarters and purchasing supplies of meat and fresh produce brought a significant infusion of cash to New Mexico. The army in New Mexico "injected comparatively large sums of money into what had been primarily a barter economy. The money was widely, if unevenly, distributed, reaching all segments of the population, including the Pueblo Indians. Not only did the army provide a market for some of New Mexico's traditional products; it created a demand for products that earlier had not been available at all or had been produced in very limited quantities."[8] The presence of the army

Figure 1.1. West side of Old Town Plaza, Albuquerque, NM, ca. 1880. Courtesy of Center for Southwest Research, University Libraries, University of New Mexico; CSWR PICT 997-001-0006.

affected Albuquerque's economy especially, since it housed an army garrison and even served occasionally as headquarters of the military Department of New Mexico.[9] A young US Attorney, William W. H. Davis, commented about Albuquerque in the early 1850s: "The army depots are located here, which causes a large amount of money to be put in circulation, and gives employment to a number of the inhabitants."[10]

With eventual profound consequences for Albuquerque and the rest of New Mexico, in the 1850s, Congress directed the US Corps of Topographical Engineers to undertake four ambitious surveys across the West. Their aim was to determine the most practicable route for a railroad from the Mississippi River to the Pacific Coast. As the leader of one of those surveys, Lt. Amiel W. Whipple wrote in his report to Congress, "Notwithstanding the richness of her mines of gold, of silver, of copper, and of iron, the deposits of coal that have been discovered in New Mexico have probably a more direct and practical bearing upon the project of a railway."[11] The summary map of the reconnaissance route prepared by Whipple depicts much of the corridor eventually followed by the main line of the Santa Fe Railway across central New Mexico. Sectional rancor followed by eruption of the monumentally destructive Civil War put the transcontinental railroad project on hold for nearly a decade after completion of the surveys.

The end of war in 1865 had further economic consequences for Albuquerque and New Mexico Territory. First, many uprooted veterans headed West, including some to New Mexico. This was especially true for Confederate veterans because the US Southeast had been devastated physically and socially by the war. As but one hint of the resulting movement, the 1870 census showed fifty-four residents of Albuquerque born outside of New Mexico.[12] Overall, Albuquerque's population grew by just under a hundred between 1860 and 1870, but more than half of that growth comprised newcomers to the region.

A factor of major importance for New Mexico's future was the post–Civil War explosion of railroad expansion westward from the Mississippi River. As part of the US military strategy during the Civil War, the Pacific Railroad Act of 1862 authorized creation of the first

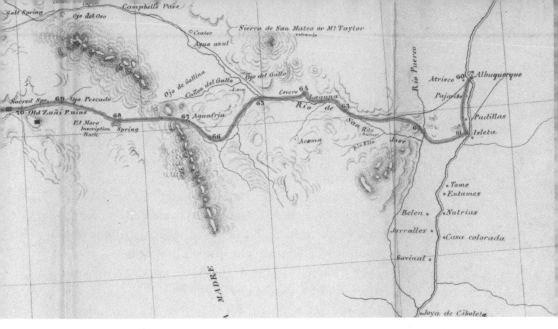

Map 1.1. Central New Mexico, detail from A. W. Whipple, Map No. 2, Reconnaissance and Survey of a Railway Route from Mississippi River . . . to Pacific Ocean . . . 1853–4. In Senate Ex. Doc., No. 78, 33d Congress, 2d Session, Vol. III. Actual survey route highlighted.

transcontinental railroad. Pivotal to Albuquerque's future, following the Civil War the US rail network expanded from the Mississippi River to the Pacific Coast based on the 1850s surveys. The immediate task was to choose among four rival corridors: one through the northern Rocky Mountains, one through the central Rockies, another skirting the southern Rockies, and a southern desert route.[13] The central Rockies route was given priority. As later amended, the 1862 act authorized special financial support and incentives to the Central Pacific and Union Pacific railroads, which would together create the first transcontinental rail line.[14]

That long line of track would result from the Union Pacific Railroad building westward from Council Bluffs, Iowa, while the Central Pacific laid track eastward from Sacramento, California. No construction started until 1863, and then only from Sacramento. That work was extremely slow, both because the Sierra Nevada presented a formidable obstacle and because most construction materials had to be sent by ship from the eastern United States via the isthmus of Panama.

Construction westward from Council Bluffs did not begin at all until the end of the war in 1865, nor did it progress far at first. There were plenty of potential laborers, with the release from military service of hundreds of thousands of men at the end of the war. "The labor competition of the eastern manufacturing plants was lessened because of the depression which immediately followed the war, and many of the soldiers went west because there was no other opening for them."[15] Construction materials, though, were another matter. Rails and other parts were urgently needed to rebuild Southern railroads that had been heavily damaged during the fighting. The result was that by the end of 1865 the Union Pacific had laid only forty miles of track. A year later, however, found the railhead at North Platte, Nebraska. One more year carried the end of track to Cheyenne, Wyoming. By the end of 1868, the Union Pacific line had passed the Green River. And on May 10, 1869, the Union Pacific and Central Pacific joined at Promontory Summit, Utah Territory.

That single string of track was never considered to be the end-all of railroads in the West. Even before the union ceremony at Promontory Summit, other railroads were pushing slowly westward from the Mississippi and eastward from the Pacific Coast. During the 1870s, constructed railroad mileage in the West swelled to 151 percent of what it had been at the end of the 1860s. The decade was one of proliferation of small rail lines running on at least six different gauges of track. As long-distance, east-west train travel became increasingly common, and especially after completion of the first transcontinental lines, it became obvious that, for convenient and accurate scheduling of long runs, there would have to be consistent standardized time zones. By 1876, a timetable convention established a version of the time zones we are now familiar with in the lower-forty-eight United States. Seven years later, in 1883, the American Railway Association adopted standard time zones for the United States. That seemingly small and obvious step brought a welcome order to railroading chaos. Likewise, the gradual widespread adoption of a 4-foot, 8 1/2-inch track gauge (straddle or distance between the inside edges of the rails) moved Western railroads toward compatibility and the possibility of national interconnectedness by rail.[16]

The extraordinary acceleration of freight and passenger movement

Figure 1.2. Celebration of completion of Transcontinental Railroad, May 10, 1869. Photo by Andrew J. Russell. Note cracked photographic plate.

across the country in the 1870s and its clear profitability stimulated an explosion of railroad building in the West. Among revitalized ventures was the Atchison and Topeka Railway, which had been founded in 1860 and focused at first on shipping Kansas wheat to eastern mills. After completion of the first transcontinental railroad, though, the Atchison and Topeka expanded its horizons, adding "Santa Fe" to its name, and slowly lengthening its line westward. By fits and starts, and after bankruptcy, reorganization, and the name change, the AT&SF had reached as far west as the Kansas–Colorado border by the end of 1872.[17] Western railroad building was given a powerful added impetus by the continuing discoveries of gold and silver deposits in Colorado throughout the 1860s and 1870s, as well as the lure of an all-weather rail route to the West Coast.

Lingering effects of the national economic panic of 1873 forced a slowdown in AT&SF construction, but gradually crews laid track across southeastern Colorado, arriving at Trinidad in 1876.[18] In this stretch, the route of the tracks generally followed the Mountain Branch of the Santa Fe Trail. As the culmination of an intense and protracted rivalry between the AT&SF (through its subsidiary the New Mexico and Southern Pacific Railroad) and the Denver and Rio Grande Railway (D&RG), a Santa Fe survey and track crew took control of key sections of Raton Pass in the frigid early morning hours of February 27, 1878. By doing so, that crew shut out the D&RG from the traditional travel route and assured AT&SF's dominance among railroads of New Mexico and much of the rest of the Southwest.[19]

All that railroad-building activity had an effect even on distant Albuquerque. The decade of the 1870s, including the final months of 1879 and the first months of 1880, during which the Atchison, Topeka, and Santa Fe Railway tracks were extended toward Albuquerque, saw an increase of more than a thousand people in Albuquerque's population.[20]

Historian Victor Westphall has offered this snapshot of business in Albuquerque in 1879, on the eve of the arrival of the rail line:

The leading merchants at this time were Franz Huning and Stover and Company. In 1870 the town had five lawyers and two doctors while now [in 1879] there were three of each. John Murphy's was still the only drug store. William Brown had dropped his advertisement as a chiropodist and dentist and was confining himself to the barber trade. He was still the only barber in town. Two blacksmith shops had been added to the one owned by Fritz Greening in 1870, while Wm. Vau and Wm. H. Ayres still had the only carpenter shop. Of bakeries there were still only two, however there were now three butcher shops instead of the one owned by Tom Post a decade before. There was still only one saloon but the merchants continued to sell liquor by the gallon. Major Werner had abandoned his hotel venture when his work as notary public and probate clerk began to take all of his time. That left two hotels owned by Tom Post and Nicholas Armijo respectively. This

Figure 1.3. Engine at Raton tunnel, 1886–1888? Photo by J. R. Riddle. Courtesy of Palace of the Governors, Photo Archives, (NMHM/DCA), negative number 038211.

was one more than there had been in 1870. A few new ventures had been started since the beginning of the decade. There was one watchmaker or mender, one tailoring establishment, and two cobblers. Such was the picture of Albuquerque on July 4, 1879, when the railroad reached Las Vegas.[21]

Attorney and historian Ralph Emerson Twitchell, who had moved to New Mexico just two years after the railroad reached Albuquerque, recalled,

> The principal event occurring during the administration of Governor [Lew] Wallace [October 1878–March 1881] was the building into New Mexico of the Atchison, Topeka, and Santa Fe Railroad. Under a charter issued by the territory to the New Mexico and Southern Pacific Railroad Company, the line crossed the Raton mountains November 30, 1878, and in February 1879, the first passenger train, carrying members of the Colorado legislature, was run to Otero station, Colfax County. In early December 1878, the Santa Fe tracks had already reached New Mexico's northern territorial boundary. The line reached Las Vegas, July 1, 1879, and was formally opened to passengers and traffic on July 7th.[22]

In his 1879 report to stockholders, Thomas Nickerson, president of the AT&SF, announced that after a lengthy negotiation, the Railway had "secured an interest in the valuable franchise of the Atlantic and Pacific Railroad company, which gives your road right of way across Arizona and California to the Pacific Coast."[23] That meant that the AT&SF could indeed become part of an integrated transcontinental system, giving it a huge advantage over its southwestern competitors.

Westphall writes, "The year 1879 drew to a close and the scene was set for some remarkably rapid action during the early months of 1880." By February 9, crews had laid track as far as Galisteo and were "moving ahead at the rate of about a mile a day." In early April the AT&SF railhead was just a couple of miles from Albuquerque.[24]

About two years earlier, the AT&SF's chief engineer and chief surveyor had already come to New Mexico on a scouting trip, "looking for the possibilities of locating a main division point somewhere on the Rio Grande."[25] The main competition was between Albuquerque and Bernalillo. A large Bernalillo landowner asked for payment of a very high price for right-of-way there, which tipped the scales in favor of Albuquerque. The Middle Rio Grande offered a major water source about halfway between Atchison, Kansas, and the California destination of

Figure 1.4. 9th Cavalry Band on the Plaza, Santa Fe, NM, 1880. Photo by Ben Wittick. Courtesy of Palace of the Governors, Photo Archives, (NMHM/DCA), negative number 050887.

the projected AT&SF Pacific line, as well as proximity to the Cerrillos coal beds. So, Albuquerque became the site of an AT&SF division point and the location for a major steam locomotive repair shop.

The official welcome of the AT&SF railroad to its future yards and the surrounding town site took place on April 22, 1880. Setting the tone for the festivities was the band of the US 9th Cavalry, an African American regiment, then on detached service at the headquarters of the District of New Mexico in Santa Fe. Leaving the plaza at the Hispanic village of Albuquerque an hour before noon, the band led "a procession to the railroad along newly designated Railroad [later Central] Avenue."[26]

Figure 1.5. Arrival of the Atchison, Topeka & Santa Fe Railroad, Albuquerque, NM, April 1880. Photo by Ben Wittick. Courtesy of Palace of the Governors, Photo Archives, (NMHM/DCA), negative number 143091.

A special passenger train carrying dignitaries from around the territory pulled into the makeshift Albuquerque station a little after midday. The out-of-town guests and a local crowd, many on horseback, then attended a program of mostly self-congratulatory speeches. The keynote of the event was struck by William Hazeldine, one of the principal promoters of and investors in the new railroad town and the station and shops complex: "When on this eventful morn the first struggling beams of light broke over the brow of yonder range of [the Sandia] mountains, grave sentinels standing guard eternally over our beloved and fertile valley, the day was born that was to be the day of all days for Albuquerque, the Queen City of the Rio Grande, a day long expected and anxiously looked forward to by the friends of progress and advancement, a day ever after to be known and remembered as that on which our ancient city of Albuquerque . . . was through the pluck, vim and enterprise of the management of the A. T. & S. F. RR. connected with the rest of the civilized world."[27]

The smoking, steaming locomotive with its string of fancy passenger cars sitting on the newly laid tracks underscored Hazeldine's words. Enthusiasm was abundant that day, and there was also an undercurrent of trepidation. But as it turned out, there was no turning back. Albuquerque would be unalterably changed by the presence of the railroad. In one fell swoop a nineteenth-century industrial city was born and already growing. As writer V. B. Price puts it, "The railroad brought with it a rush of newness, a revolutionary alteration in the rate of change, a 'future shock' that has hit the valley in shock wave after shock wave ever since."[28]

CHAPTER 2

The Requirements of Steam

The revolution in transportation in New Mexico and all of the West brought about by the introduction of steam locomotion in the last decades of the nineteenth century was dramatic and immediate. Most obviously, the speed of travel increased by an order of magnitude. In the 1840s a hurried wagon trip from Santa Fe to Kansas City–Independence, Missouri, was "usually made in about forty days."[1] In 1880, by contrast, Adolph Bandelier, riding an AT&SF train, covered that same distance in the opposite direction in less than *fifty-eight hours*, just over two days.[2]

Such speed was possible only with constant maintenance of tracks and lubrication, watering, refueling, inspection, and frequent overhaul of the steam locomotives that pulled the cars full of passengers and freight. Steam locomotives were particularly vulnerable to catastrophic disaster because they combined active, energetic fires on board; superheated steam circulating through long, serpentine arrays of pipes; and hundreds of rapidly moving parts meshed with others; all moving together as a unit at significant speed along iron rails that were constantly being shaken and jarred out of alignment.

The most dangerous single component of a steam locomotive was the boiler and its associated steam conduits (flues). As a 1930 film record of locomotive repair made clear, "the life of a [railroad steam] engine is largely dependent on the life of its boiler." Natural impurities in the water that circulated through boilers could precipitate out as scaly deposits that would slowly constrict passageways, dangerously raising pressure in the system. Some impurities could cause frothing or scale could break loose. Either of those eventualities could suddenly clog pipes, raising pressure in the whole steam system beyond the strength of even iron and steel. Nineteenth-century newspapers regularly reported steam locomotive explosions that would destroy rolling stock and nearby buildings, and kill or grievously maim railroad workers. Typical is this report from April 1904: "A locomotive boiler exploded, on May 15th, on the Santa Fe Railroad, near Bagdad, Cal. Engineer S. Ebbutt received injuries from which he died shortly afterwards. Fireman J. F. Showalter also received minor injuries."[3] As early as 1866, the Hartford Steamboiler Inspection and Insurance Company was incorporated at Hartford, Connecticut. One of its avowed goals was to inspect steam boilers and insure the owners against loss or damage arising from boiler explosions. The efforts of this and similar companies led to adoption by railroads, including the AT&SF, of routine, detailed locomotive boiler inspections undertaken at repair shops.

To actually move a steam locomotive, the high-pressure steam was admitted alternately to one side and then the other of reciprocating pistons. The level of the water in the boiler had to be carefully watched by the engineer and fireman. Their judgment was only as good as the combination of gauge cocks and water glasses that constantly showed the water level in the boiler. Periodic testing, cleaning, and repair of the gauge cocks was literally of life-saving importance and required keen judgment on the part of the locomotive crew.[4] The quality of water sources for use in boilers varied through time. In New Mexico and many other places, during the late nineteenth and early twentieth centuries, unprocessed or imperfectly processed ground water (including spring water) was regularly used in locomotive boilers. Thus, cleaning "mud" and scale from boilers was a crucial task performed at the Shops. Furthermore, the effectiveness of steam in moving the pistons depended on how thoroughly the piston rings sealed against the interior of the steam cylinders. Therefore, the efficiency and power of the steam engine hinged on periodic replacement of the piston rings, as well as hundreds of other parts.

Figure 2.1. Sectioned fire-tube locomotive boiler and firebox from a DRB Class 50 Locomotive, dating between 1939 and 1948. Photo by Rabensteiner. GNU Free Documentation License at fsf.org.

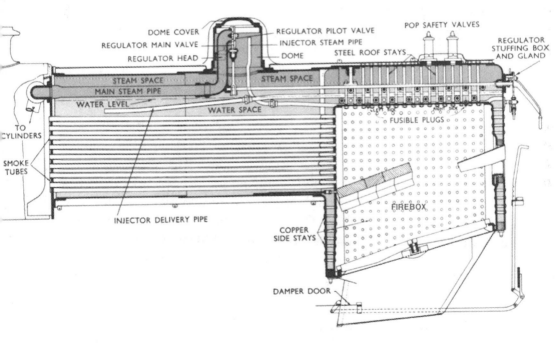

Figure 2.2. Diagrammatic cross-section through an early steam locomotive boiler and firebox. From *Handbook of Steam Locomotive Enginemen* (British Transport Commission, 1957) (label of firebox added).

The pistons articulated with a complex of dozens of rods and levers that turned the large drive wheels. Precision-ground and polished bearings (or journals) and shafts were located at each point of connection in the resulting power train to minimize friction. Grease, oil, and other lubricants reduced friction at the bearings even further. But through time the rotation of wheels and cranks wore away the surfaces of bearings and shafts. Without periodic machining or replacement, what had once been a smoothly running assembly would eventually seize up and stop a locomotive dead. So all of those parts had to be retooled or replaced periodically. According to Chris Wilson, "Over the fifteen-year life of an average locomotive, it might be rebuilt or receive other major shop repairs once every 12 to 18 months."[5]

In his book *Iron Horses: America's Race to Bring the Railroads West*, Walter Borneman writes, "The locomotive was the beating heart of a train, but brakes were the circulatory system that allowed it to function."[6] In the AT&SF's early days, "individual cars were outfitted with brake wheels at one end. When these wheels were tightened, the connecting rigging clamped brake shoes against the running wheels . . . creat[ing] enough friction to slow their revolutions, thus slowing the trains. . . . Operating these brake wheels gave rise to the most dangerous job in railroading. At a whistle signal from the engineer, nimble brakemen ran atop the cars—jumping from one swaying car to another—and frantically set the brakes."[7] During the 1870s, air brakes became common on trains, utilizing long compressed-air lines running the length of a train. By the mid-1880s, though, George Westinghouse deployed a more reliable system with an independent air cylinder on each car. With this system, if the main compressor failed or cars broke free of a train, "the brakes would set automatically and in theory stop the train."[8] As fail-safe as automatic brake systems seemed to be, they were only as effective as the most recent replacement of brake shoes and servicing of air cylinders.

No matter the source of a locomotive's power, its ability to run smoothly along the rails with a minimum of risk of derailing depended on the condition of its wheels, both driven and rolling. Especially critical were the flanges that held the wheels against the rails, the source of

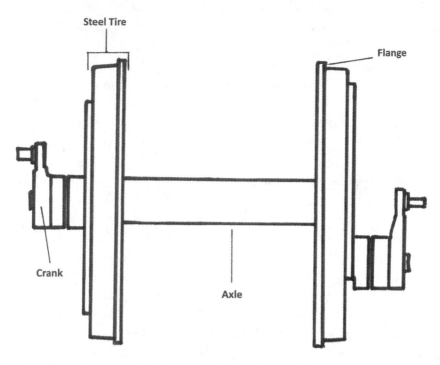

Steel Tire

Flange

Crank

Axle

Figure 2.3. Railroad wheel showing flange and steel tire. Author's drawing.

much of the squealing that one hears as a train passes by. And flanges are only the most obvious element of wheels that determined safe and efficient operation of locomotives. There is a large number of other wheel parts that played a critical role in moving a locomotive and its train of cars, including suspension systems, axles, journals and bearings, and swivels. The malfunction of any of those parts could render a locomotive inoperable. And even with all wheels turning smoothly the locomotive could lose traction in rainy, snowy, or icy weather. To mitigate against that, steam locomotives were equipped with sand domes and pipes that dribbled sand onto the rails immediately ahead of the wheels. If, under adverse conditions, the delivery of sand were interrupted, a locomotive could come to a standstill with wheels spinning, especially on a grade.

These are only the most critical physical components of steam railroading, all of which were inspected, serviced, and maintained during scheduled maintenance at shops like the ones built at Albuquerque.

Without routine repair and overhaul of literally thousands of parts, steam locomotion would soon come to a halt. For many decades steam locomotives and steam marine engines drove the vast majority of transportation in the United States and around the world. If steam engines didn't run because of a lack of parts or skilled machinists, neither did the nation's trade and commerce.

Beyond essential maintenance and repair, there were also two vital supplies that kept steam locomotives running for about a hundred years: fuel and water. Coal was the fuel of preference for most of the time steam locomotives dominated railroading. That was because it was denser and could therefore generate more heat than wood, and it took up less space to carry, which meant fewer fueling stops were necessary. Albuquerque and New Mexico presented several sources of coal that were used for steam locomotives. Large coal fields existed in relative proximity to the AT&SF rail line near Gallup and Raton, with a smaller but rich field in the Cerrillos area. "As much as 45,000 tons of anthracite [coal] were mined annually from the Cerrillos field during the period 1888 to 1957."[9] Not coincidentally, that period is almost identical to the time span of coal-powered steam locomotion in New Mexico.

The Middle Rio Grande Valley at Albuquerque offered abundant, shallow ground water.[10] Like nearly all water used by steam railroads, water in New Mexico had to be treated in the tender in order to minimize the deposition of calcium carbonate scale within locomotive boilers. A tender was a specialized car coupled to a locomotive that carried the supply of water and coal or other fuel.

Both coal and ground water were also necessary at every stage of locomotive repair and overhaul. Hot water and piped steam were in use every day, all day throughout the Shops for cleaning parts, tools, clothing, shop areas, and shopmen. In the early days of the Shops, stationary steam engines supplied direct power for industrial-scale tools, such as power hammers in the blacksmith shop, via a system of belts. And even after conversion of most tools to electricity, a steam-driven power plant generated that electricity. All of those functions ran on water converted to steam mainly by the burning of coal.

CHAPTER 3

Albuquerque and the Locomotive Repair Shops in the Early 1880s

In order to erect the various buildings that would comprise the Albuquerque station and shops complex, AT&SF had to first acquire the land on which they would be built. American railroads had recourse to several different means of land acquisition, ranging from very generous grants from the federal government to purchase to outright theft. Central New Mexico in 1880 was a region already organized according to long-standing land tenure rules, many inherited from the time of Spanish and Mexican authority over the region. Thus, AT&SF expected to have to purchase the land where its Albuquerque facilities would be located.

A similar situation existed concerning residences for the hundreds of anticipated locomotive-shop employees. The area where the Shops and their surrounding halo of residences were likely to be built was already known as Barelas. It was a small village consisting of houses and other buildings associated with farm fields that occupied the floodplain on the east side of the Rio Grande a little less than two miles southeast of the Albuquerque plaza, now known as Old Town Albuquerque.

At a meeting of residents held on July 8, 1879, discussion centered on the possibility of the town donating land to the railroad as an incentive for the AT&SF to formally designate Albuquerque as a division point, which would bring a massive economic infusion to the town. Father Donato Gasparri, head of a group of Jesuits who had recently arrived in New Mexico, maintained that "poor people could not afford to donate their scant property to the railroad company, that the wealthier members of the community should contribute toward the purchase of these lands from the poorer people, and that the grant could then be made to the railroad company."[1]

By the last days of 1879, "the advance guard of the railroad—laborers, speculators, traders, contractors, etc.—had already come to town, and it was expected that the line would be completed by the fifteenth of March of the following year."[2] In keeping with the Gasparri proposal, enterprising Albuquerque residents, especially a handful of recent newcomers, sought to buy up swaths of land in the vicinity of Barelas, through which the rail line would presumably run and where the main support infrastructure would have to be built. Among those entrepreneurs were Franz Huning, William Hazeldine, Elias Stover, Santiago Baca, and John Phelan.

Even weeks before the railroad reached Albuquerque, it

was now a busy little town indeed. The Central Bank had been organized with Jefferson Reynolds as president, and instead of one saloon there were now fifteen. There were two hardware stores, a saddlery, a shoemaker's shop, two Chinese laundries, six architects and builders, about twenty carpenters, two seamstresses, two pawnbrokers, two wholesale liquor stores, a planing mill, a grist mill, two drug stores, half a dozen restaurants, a tan yard and wool pulling house, a sash door and blind store, and the professions were represented by five doctors, six lawyers, one assayer, and one editor.[3]

In March and early April, 1880, [Huning, Hazeldine, and Stover] were furiously buying up land between Barelas Road and the proposed depot site. . . . This was the area later to be known as the *Original Town Site*. It seems certain that these Albuquerque citizens were acting under the auspices of the New Mexico Town Company [an organ of the AT&SF] . . . which was organized March 3, 1880. . . . The land which comprised the actual depot grounds was purchased by the three between March 3 and April 3. . . . On April 9 and 10, Huning, Hazeldine and Stover deeded their holdings to the N[ew] M[exico] and S[outhern] P[acific] [wholly owned by the AT&SF] for $1.00. Furthermore, on May 8, they deeded the whole of the original *Town Site* to the New Mexico Town Company, likewise, for $1.00.[4]

Other deals that facilitated establishment of the Albuquerque Locomotive Repair Shops included one in June 1880, between Franz and Ernestine Huning and the Atlantic and Pacific Railroad (A&P, soon to be another subsidiary of the AT&SF) by which the Hunings transferred somewhat less than an acre lying just north of the A&P's station grounds to the railroad for one dollar.[5]

Two Barelas area farmers, Antonio Candelaria and Ignacio López, refused to sell to Huning and his partners, but later sold their properties directly to the railroad. With the railroad tracks laid down between the main Barelas irrigation ditch and the fields that it had irrigated, farming became impossible there.[6] Huning, Hazeldine, and Stover signed an agreement with the railroad that "they were to receive jointly from the [New Mexico Town] Company one-half of all net profits derived from the sale of lots situated on land owned by [the] Company."[7] By this means, as well as through direct sales of lots, the three men were to gain fortunes selling land they had just themselves purchased in the new Barelas and San José neighborhoods. Their customers would be largely the families of AT&SF employees at the Locomotive Repair Shops (see appendix 1).

A February 1881 contract lists 298 lots in the Atlantic and Pacific Addition across 1st Street from the railway station and Shop grounds that were owned and offered for sale by Franz Huning, Elias Stover, and William Hazeldine.[8] The asking price for those lots ranged from a low of $75 (for a 50 x 142-foot lot at the corner of Broadway and Iron Avenue) to a high of $450 (for a 25 x 175-foot lot on Atlantic Avenue between Barelas Road. and 5th Street). At those rates, the total asking price of the 298 lots was well over $25,000.[9] In today's dollars that would be the equivalent of more than $550,000.[10]

On the east side of the tracks from the Atlantic and Pacific Addition, Huning and John Phelan laid out the Highland Addition, comprised of 34 blocks made up of 350 lots, including some of irregular size and shape. One of the early purchasers in the Highland Addition was Peter Quier, who bought lot 10 in block 9, on Broadway between Railroad/ Central Avenue and Gold Avenue for $100 in December 1880 (see map 3.1). In 1896, a relative of Peter Quier's, presumably a son, was still residing on that property, and other relatives were living on the Arno Street side of the same block.[11]

Within weeks of the opening of the rail line as far as Albuquerque in April 1880, numerous construction projects were underway in Barelas and San José: a station house, a roundhouse and coaling and watering facilities, as well as residences for railroad workers. A civil engineer, Walter Marmon, who had been living at Laguna Pueblo for several years, laid out the streets and blocks in what were to become the neighborhoods of Barelas and South Broadway. Lots, though, were slow to sell at first, even at prices as low as ten dollars each. A number of the first buildings that were erected were prefabricated structures carried by the railroad in pieces on flatcars from one location to the next, as the end of the line moved south and west. The railroad depot that welcomed the first passenger train to Albuquerque was an assemblage of old boxcars.[12]

But it wasn't long before more substantial homes became the norm.

Housing tracts platted on the east side of the railroad tracks . . . included arguably the city's most prominent early neighborhood: the Huning Highlands Addition that was developed by. . . . Franz Huning. Touted as 'Albuquerque's first subdivision,' residential housing in Huning Highlands Addition began in 1880 and within eight years sixty-three percent of its 536 lots had been sold. It was platted at the base of the sand hills just east of Broadway and spanned both sides of Railroad [Central] Ave. It was thus situated slightly above the lower-lying valley which was marked by the feverish noise and activity accompanying the industrial and commercial development occurring across the railroad tracks.[13]

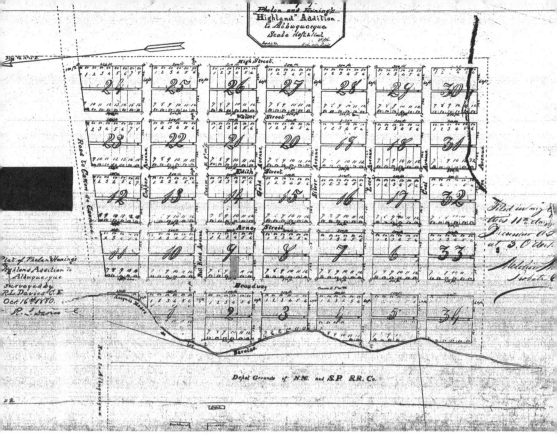

Map 3.1. Map of John Phelan and Franz Huning's Highland Addition, December 1880. Lot 10, Block 9, Purchased by Peter Quier, Highlighted. Line of the New Mexico and Southern Pacific Railroad (AT&SF) shown at the bottom. Bernalillo County Clerk's Office.

This included most notably the Albuquerque Locomotive Repair Shops.

> In large part because of its 'boom town' beginnings, and perhaps because of [New Town Albuquerque's] somewhat transient population initially, single-family houses . . . were supplemented by multi-unit dwellings. Within ten years of the town's founding, there were thirteen hotels, lodging houses, and houses offering 'furnished rooms.' . . . These included the city's multi-story hotels such as the Armijo House and San Felipe Hotel as well as more modest two-story wood frame structures such as Strong's European Hotel . . . and the Windsor Hotel. . . . The latter were both popular with railroad workers.[14]

Even with only vestigial locomotive repair shops at Albuquerque and residential areas for anticipated employees barely platted out, the work of extending the tracks farther west went forward. In October 1880, *Railway Age* reported that AT&SF "has just received twenty-eight boarding cars [dormitory cars], which will enable it to put on a still larger [work]force. Thirty-five miles of iron [track] are now in the Albuquerque yards, and material for 100 miles more is on its way. The contracts for grading as far as Fort Wingate are about to be let."[15] New Albuquerque hardly had time to become more than a name on a map before road crews were also working farther south along the Rio Grande, pursuing the quickest connection to California.

According to Ralph Emerson Twitchell, "Following down the Rio Grande valley, [the AT&SF] was completed to Deming [NM], March 10, 1881, where a connection with the Southern Pacific Railroad was made, thus forming the first all-rail route across New Mexico to San Francisco." With construction of that stretch of track from Albuquerque to Deming, the second transcontinental railroad in the United States became a reality, and Albuquerque was truly launched as a transportation hub.[16]

That route, though, was a circuitous one for passengers and goods bound for San Francisco, which at the time was the major destination in California. Even with the Deming route in operation, AT&SF pushed its A&P track westward, loosely following the thirty-fifth parallel. By January 1883, the AT&SF/A&P track had reached to within 110 miles of crossing the Colorado River into California, "leaving only 130 miles to be completed to secure another connection between the Atlantic and Pacific Oceans."[17] Less than two months later, the line was only sixty miles from California.[18] And in July it was announced that "track on the Atlantic & Pacific railroad is now finished to a connection with the Southern Pacific of California at the Needles, completing another great through line to the Pacific which will be opened as soon as the bridge across the Colorado River is finished."[19] In the meantime, AT&SF had officially absorbed the Atlantic and Pacific so that as a single railroad it now stretched from Kansas City to Needles, connecting at both ends with other lines to complete the transcontinental route.

Figure 3.1. Railroad (Central) Avenue at 1st Street, 1881. Photo by Cobb Studio. Courtesy of Center for Southwest Research, University Libraries, University of New Mexico; CSWR PICT 000-119-0575.

Despite the optimistic plans of 1880, the new railroad town of Albuquerque did not immediately mushroom into a little metropolis. Some of the people who, in the first wave of enthusiasm, began building near the tracks, gave up and moved before their structures were finished. By the end of the year 1880, however, the newly platted town became the locus of a genuine building boom "that was to continue unabated for years to come."[20] Within a matter of only a few years, both residents and visitors were astounded by the transformation. C. M. Chase, a newspaper editor from Vermont, for example, noted that "last February [1882] in the locality of the depot there was nothing but two or three shanties and a few cloth tents." In the following year, though, "Railroad [Central] Avenue was nearly solid with business houses for five blocks," and other buildings, both commercial and residential, were multiplying with astonishing speed. Gone were the days of cheap building lots. Now lots were selling for between $200 and $2,000 each.[21]

Remembering New Albuquerque after it had existed for just three years, Sylvester Baxter wrote:

To the rapid growth of the place I can testify. Returning after a month's absence in 1881, I found that the number of buildings had about doubled. The manufacture of adobes was going on at a prodigious rate, and there was a lively clatter of carpentry in the erection of frame buildings. . . . Visiting Albuquerque again a year and a half later, in 1882, I found the changes that had taken place in the mean time still more remarkable. Where at that time there was but one business street, lined with an inferior class of buildings, and scattered houses dotted here and there over the level fields, outlining the anatomy of the town that was to be, the skeleton had become clothed with good solid urban flesh, or, to speak more liter-ally, with brick, stone, adobe, and timber. The buildings now stood in sturdy ranks. Railroad [Central] Avenue had been paralleled by another and a handsomer business street named Gold Avenue; the intersecting cross streets had also been built up with business houses; large and glittering plate-glass windows were filled with attractive goods in the latest fashions. . . . the streets were brightly illuminated by a gas of excellent quality made from coal mined out on the Atlantic and Pacific Railway near the Arizona line. . . . The first brick had been manufactured in the town only a few months before, and there were already numerous brick buildings of sub-stantial architecture on the business streets.[22]

Most important for the aspiring community's future, "The first [locomotive repair] shops were built at Albuquerque in 1881. . . . The more important buildings erected at that time consisted of a machine shop, a boiler shop and a blacksmith shop, all of which were con-structed of red sandstone walls" with timber roof framing and tile roofs.[23] By 1886, a bird's-eye-view map of Albuquerque clearly depicted two large locomotive shop buildings (the machine and boiler shops), an associated power plant, and a roundhouse. Years later, when the old machine shop was about to be demolished, an accompanying scaled drawing of it was made.

Evidently, even before the tracks reached Albuquerque, perhaps as early as 1879, the Santa Fe had at least one water well dug at the

Figure 3.2. Detail from "Bird's eye view of Albuquerque, Bernalillo Co[unty], New Mex[ico], 1886," map by Augustus Koch, showing buildings of the Locomotive Repair Shops (#17). Courtesy of Center for Southwest Research, University Libraries, University of New Mexico; CSWR G4324.A4A3 1973 .K6 c.1.

Figure 3.3. Photo showing the old machine shop, taken from the southeast before its demolition, 1922? Photographer unknown. Courtesy of Center for Southwest Research, University Libraries, University of New Mexico; Albuquerque Const. Sites Album, CSWR PICT 2002-013-0033a.

location of the future shops and erected a windmill to pump water from the well, in addition to a 30,000-gallon water tank. By 1888, the Shops were already being supplied from six of an eventual fourteen water wells.[24] Abundant water was an essential resource at any steam locomotive shop site, not only to supply the locomotives but also because all the machinery and power tools ran either directly by means of belts from stationary steam engines or via steam-generated electricity.

Population exploded as well. Lina Browne writes, "Some five or six years after the arrival of the railroad [that is, in the mid-1880s] the town of Albuquerque had a population of 6,000, more than four times larger than [the old town] had been in 1860."[25] Industrial labor at the Repair Shops and associated businesses was the main driver of the transformation of Albuquerque. Because of that dominance by urban industrialism, Albuquerque was unlike any of the other towns in the territory. There were linkages and similarities with other railroad towns in New Mexico (Raton, Las Vegas, Los Lunas, Grants, Gallup, Deming, and Lordsburg), but there was no other railroad facility in New Mexico that came close to matching the Albuquerque Locomotive Repair Shops in size or complexity. The Shops gave Albuquerque a decided industrial character and infrastructure, which tended to attract related and similar enterprises. The concentration of industrial businesses was a self-reinforcing condition. Industries attracted more industries and further differentiated Albuquerque from other places. More manufacturing and industrial businesses meant more employment opportunities, which attracted more residents. As a result, New Town Albuquerque soon far outstripped all other New Mexico communities in size.

The residential and commercial district surrounding the AT&SF Shops was the epicenter of that growth. To the west of the shops was the Barelas neighborhood, and to the east was the San José neighborhood, which has subsequently been subsumed into the South Broadway neighborhood. As described in 2013, "Along the eastern edge of the Rail Yards is the South Broadway neighborhood. Much of the community's growth took place between 1885 and 1925, following its founding by Antonio Sandoval, a wealthy landowner responsible for constructing the Barelas ditch, which drained and irrigated the sur-rounding area. As in Barelas, many of South Broadway's residents made their living through agricultural pursuits before transitioning to jobs at the Rail Yards and local iron foundry."[26]

Because of the lack of machinists, boilermakers, and other skilled industrial laborers living in Albuquerque and New Mexico in general before the arrival of the railroad in 1880, the AT&SF Shops relied heavily on workers coming from Kansas and even farther east. There, railroads had already been operating for decades, and a journeyman workforce had developed to meet the labor demands of various railroad companies. Furthermore, steam power had been used in the East for generations in many applications in addition to railroading, from residential and commercial heating to running industrial and agricultural machinery of all kinds. Thus, there were boilermakers, machinists, and mechanics with many specialties and at all levels of experience, from apprentices to senior journeymen with decades of know-how. As a result, many of the first residents of the Barelas and San José neighborhoods were AT&SF / A&P employees who were natives of the East.

Very soon, though, the railroad's need for workers outstripped the supply of trained specialists from farther east. Although the railroad's preference was clearly for experienced shopmen, which generally meant newcomers to New Mexico, AT&SF / A&P and other railroads had difficulty recruiting workers to move to the Southwest. "In 1891 A&P General Manager Robinson complained that he could not get first-rate railroaders because no one who could find work anywhere else would settle in New Mexico or Arizona."[27] That led to employment opportunities for some New Mexico natives, who in other circumstances might not have been hired. There were also some New Mexicans who had experience as blacksmiths and other sorts of metal fabricators, as well as upholsterers and wagon and carriage manufacturers, which made them desirable as employees in the Albuquerque Locomotive Repair Shops.

In great part to facilitate the employment at the Repair Shops of local New Mexico natives, the Street Railway Company was incorporated in 1880 and "by the end of the first year, it had eight mule-drawn cars and three miles of track connecting the Old Town plaza with 'New Town' and the suburb of Barelas."[28] Hannah Wolberg describes the Albuquerque streetcar system:

> The co-founder and president of the streetcar company was New Yorker Oliver E. Cromwell, who was [already] an investor in Albuquerque real estate. He partnered up with Franz Huning and William Hazeldine to create the company. The tracks ran down the center of Railroad Avenue, now Central Avenue, from New Town to Old Town on a narrow gauge track, connecting the two sites that had become alienated from each other. In the 1890s the mules were replaced by a horse with a bell hung around its neck that alerted patrons to its arrival. Although slow, cumbersome, and sometimes dangerous due to its inclination to coming off its tracks in high winds or when the load was unbalanced, the trolley transported workers every morning and evening to and from work for ten cents.[29]

Over the years the trolley line expanded its service area. "At its fullest extent, the streetcar line consisted of approximately six miles of track from Old Town in the west to the University of New Mexico in the east; and the American Lumber Company in the north, and the A. T. & S. F. rail yards and Barelas neighborhood lay at the south end. Another spur off of Railroad Avenue ran for twelve blocks to the south along Edith Street in the Huning's Highland Addition."[30] The streetcars remained a transportation mainstay for employees at the Repair Shops until they were replaced by a fleet of gasoline-powered buses in 1928.[31] During the 2010s, a road construction crew uncovered lengths of track from the trolley, buried under Central Avenue.[32] Those sections of track are currently curated at the WHEELS Museum.

Although it did not immediately affect employment at the Albuquerque shops, an unusual arrangement between AT&SF and the Pueblo of Laguna brought a significant number of members of that tribe into railroad employment. Known as the "Watering the Flower" agreement, this oral compact was entered into in 1880 and remained in force until

1963.[33] In exchange for passage across Laguna Pueblo land and the use of water for its steam locomotives, the railroad promised that "it would forever employ as many of the Lagunas to help build and maintain the system as wished to work."[34] Initially, Laguna people worked primarily in track laying and maintenance, but eventually they held jobs throughout the Santa Fe system, including skilled positions such as machinists at the Albuquerque Shops.

By 1893, New Albuquerque had become a fairly compact town centered around the Locomotive Repair Shops and the railroad depot. "Most development [had] occurred within two blocks of the train station mostly west of the station, one block north and three blocks south."[35] Railroad-related businesses naturally also took root adjacent to the repair facilities and the freight and passenger depots. Such enterprises included the Albuquerque Foundry and Machine Works, located on the east side of the tracks, opposite and just a stone's throw from the Locomotive Repair Shops.

One result of the large number of single men working at the shops and on the trains was that they helped support "seven houses of prostitution and an opium den that lined Railroad Avenue west of Fourth Street" in 1882, according B. F. Saunders, then editor of the *Evening Review*, one of two daily Albuquerque newspapers.[36] With an altogether different influence on Albuquerque and New Mexico as a whole, in February 1889 the State Legislature authorized the University of New Mexico to be established in Albuquerque.[37]

The official census of Albuquerque in 1880 was 2,315, by 1890 it was 3,785, and in 1900 it was 6,238.[38] A significant share of the increase directly reflected AT&SF shopmen and their families. The official census figures are much smaller than the population numbers from the 1880s reported by Lina Ferguson Browne and quoted earlier. New Mexico governor L. Bradford Prince complained that the 1890 census of the state represented a serious undercount as a result of enumerators making little effort to obtain information from citizens who spoke Spanish.[39] Governor Edmund Ross predicted overly optimistically in 1887 that Albuquerque would soon be home to at least 100,000 residents.[40] Between 1870 and 1890 the Territory of New Mexico's population jumped from less than 91,000 to more than 140,000, with more than half of that increase coming from in-migration.[41] The influx of new residents, however, was not sufficient to fill the full range of jobs involved in maintenance and repair of the AT&SF's burgeoning inventory of steam engines and passenger and freight cars, which was the task of the Albuquerque Shops.

Information provided by the *Abstract of the Twelfth Census* shows the number of employees in "manufacturing and mechanical pursuits," which included "steam railroad employees," in New Mexico in 1890 as 849. A sizeable share of those must have worked at the Albuquerque Locomotive Repair Shops, but we do not have an accurate count. We can say with assurance, though, that by 1900 the payroll at the Shops was at least 501.[42]

Railway Age reported in 1883, "The machine shops at Albuquerque are large enough for all present business, and built upon a plan allowing of enlargement to any extent to meet future wants and growth of the road."[43] In less than thirty years after that rosy assertion, however, the number and size of steam locomotives proved to be too large for the original Shops, and the entire shop complex had to be replaced.

CHAPTER 4

The Railroad's Immediate and Lasting Impact

The 1880s and 1890s

Within a period of about six months, from roughly December 1879 through May 1880, the New Mexico and Pacific and the Atlantic and Pacific Railroads, both subsidiaries of the AT&SF, with the collaboration of a small group of New Mexico residents, created not only a railroad service facility on the Rio Grande but also an entirely new town. In the process, the railroad appropriated the name, identity, and much of the vitality of the old farming community of Albuquerque and almost overnight replaced it, lock, stock, and barrel, with a different community two miles away. New Albuquerque, as it soon came to be known, was organized on very different principles and for very different purposes than Old Albuquerque.

There was a new community layout and new architecture; there were new people speaking unfamiliar languages; there were new customs and religions; there was a new way and rhythm of life that its adherents touted as superior to New Mexico's venerable seasonal round. An urban-industrial ethos and way of life shouldered into and against a long-established rural-agrarian world. The result was the equivalent of a cultural tsunami that precipitated violent conflict between representatives of the two distinct populations, which lasted for years.

About this hostility, attorney and historian William Keleher wrote that by the late 1890s,

fighting and feuding between "New Town" and "Old Town" which
had been waged between the years 1880 and 1893 as to which
community was entitled to the use of the name of Albuquerque had
begun to subside, and the diehards in both camps had indicated
a willingness to bow to the inevitable and accept the fact that the
"New Albuquerque," built up around the railway depot and tracks
was to be a permanent community, separate and apart from the old
town of Albuquerque, located two miles to the west. The squabble
had been settled as between the new town and the old town over
the right to use the postmark "Albuquerque" in the respective post
offices. A ruling from Washington provided that neither one should
use the word "Albuquerque" [alone].[1]

Although Keleher portrayed this period of intense discord between
the resident Hispanic population of the Albuquerque area and the newcomers associated with the railroad as quibbling, it was in fact far from
a trifling matter. Rather, the period from 1880 until at least the middle
1890s was characterized by frequent, intense outbreaks of rage on both
sides, focused on a large range of issues, for which the use of the name
Albuquerque often stood as a proxy and a rallying symbol. For example, disparity between the two groups in terms of cash income quickly
mushroomed as many newcomers secured good-paying jobs, for the
time and place, as machinists and boilermakers in the Locomotive
Shops, while the relatively few Hispanos who were hired by the railroad served almost entirely in low-paid, low-skill positions such as
helpers and laborers. This was the result of both a higher level of
learned technical skill among the new arrivals and a pervasive anti-
Hispanic bias in the US railroad culture of the day.

The virulence of attitudes on both sides is reflected in an obviously—and today offensively—partisan summary of the animosity
between residents of Old Albuquerque and New Albuquerque written
years later by historian Benjamin Read: "Just then [during 1880] the
A. T. & S. F. system was first building its railroad toward New Mexico to bring civilization and communication with the East. Unfortu-
nately these advantages were to be accompanied by Protestant

bigotry."[2] Likewise, many of the non-Catholic railroad workers sneered at local Hispanic adherence to hierarchical religious guidance emanating from Rome.

In the 1880s and 1890s, matters of religion were only one festering sore spot between "railroad people" and "locals." Even such things as styles of clothing, habitual gestures, manner of walking, and formality or familiarity of address carried antagonistic cultural messages. The ten-hour-day, six-day-a-week railroad work schedule was sorely at odds with farming's partition between night and day, rest and labor, as well as with the routine of religious activities on Sundays and many other special days. As almost always in cases of cultural collision, there was much concern among the established Hispanic population about gender relations, especially among young people. The upshot was that there was frequent cause for friction and animosity.

Ill will between the native Nuevo Mexicanos and newcomers from the East was expressed in many ways, including sabotage of the mule-drawn trolley line between Old Town and New Albuquerque. The trolley drivers "were often harassed by boys who placed small obstacles on the tracks, or who jumped on the sides of the fragile cars and rocked them off the tracks."[3] Such vandalism was common despite the fact that it interfered with the ability of Hispanic Locomotive Repair Shop employees to get to work or to get there on time.

Long-running friction also arose between the railroad and indigenous Pueblo communities in the Middle Rio Grande Valley, which became a regular feature of life for decades. The situation at Santo Domingo Pueblo, north of Albuquerque, is well documented. In 1880, when the railroad grading and construction crews reached the eastern boundary of Santo Domingo's land grant, "a small army of construction crews drove their way through pastures, farmland, and ditches, passed within two hundred yards of the village itself, and continued south out of Santo Domingo's land grant with no advance notice, negotiation, lease, permission, purchase, or papers."[4] The railroad established a station on Santo Domingo land just two miles from the pueblo village, around which a town named Wallace quickly grew. Settlers at Wallace were soon occupying Santo Domingo farmlands

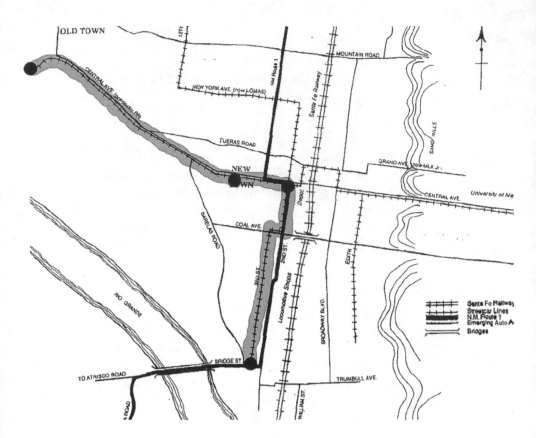

Map 4.1. Partial map of Albuquerque in 1918 showing the Locomotive Repair Shops. Original street-car line highlighted. Adapted from Tita Berger and Adam Sullins, comps., *Pedestrians, Streetcars and Courtyard Housing: Past and Future Albuquerques* (Albuquerque: University of New Mexico, School of Architecture and Planning, 2008), 13.

for their own purposes. Pueblo women were assaulted at Wallace, and Pueblo people and livestock were killed by trains that traversed the reservation.

One of the Pueblo responses to the railroad-related deaths, injuries, and thefts was destruction of railroad infrastructure: breaking telegraph line insulators, tearing down the wires, and stealing ties and poles meant for laying track and building trestles and other structures. In response, settlers at Wallace threatened to invade the village of Santo Domingo and shoot people wholesale.[5]

As skirmishes between local Native Americans, Hispanic New

Mexicans, and recent immigrant railroad employees repeatedly flared, though, AT&SF management remained focused on capturing and then expanding trade and passenger service between the eastern United States and California, which was seen as the railroad's real prize. As a result, the roster of employees at the Albuquerque Locomotive Repair Shops continued to grow, as did New Albuquerque and the Territory of New Mexico. According to the official census, between 1880 and 1900 the population of New Mexico as a whole grew by more than 63 percent, due heavily to immigration from the eastern United States.[6]

The *Albuquerque Weekly Journal* noted on December 1, 1882, that "real estate took a big boom at just about this time last year, and history seems to be repeating itself." The same issue reported, "Last evening's express on the Atlantic & Pacific brought an unusually large number of passengers." Furthermore, "Plans are now being draughted for a brick store for Ed. Strasburg, to be built on Lead avenue, near Third street." The newspaper, operated by newcomers, was full of glowing and optimistic stories such as this one regarding the building of a very expensive house: "J. M. Wheelock is preparing the plans for a residence for J. N. Scott, of Chihuahua. The estimated cost of the structure is $15,000. Mr. Wheelock is also busily engaged at work on the plans for a convent to be erected in the Armijo Bros. addition. The plans show that the building will be a beauty." In addition, the classified ads in the same edition of the newspaper listed eleven attorneys and seven physicians in New Albuquerque.[7] All of this confirms that the new town was now expanding rapidly.

Writing about this early period of New Albuquerque's history, attorney and historian Ralph Emerson Twitchell later stated,

New Mexico was then [at the end of Lew Wallace's governorship] enjoying a great period of prosperity. The livestock, mining, and other industries were making tremendous strides. Investments and speculation in land grant holdings, town and city lots in all the cities and towns along the new railroad lines, and in other collateral enterprises, marked the prosperity of the period.[8]

Historian William Dodge notes that during the last two decades of the nineteenth century,

Manufacturing businesses in downtown Albuquerque benefitted from the arrival of the railroad. The Southwest Brewery and Ice Company produced 30,000 barrels of Glorieta Beer per year and forty-five tons of ice per day in its factory situated on the east side of the tracks at Roma Ave. The Albuquerque Foundry and Machine Works, located east of the tracks across from the rail yards, manufactured as much as $100,000 worth of iron goods per year—much of it for the AT&SF. Albuquerque's central location in the territory bolstered its potential as a warehousing and distribution center. Immediately upon the railroad's arrival, large commercial warehouses, such as the Charles Ilfeld Wholesale Company and the Gross, Kelly and Company, appeared on either side of the tracks, both north and south of Railroad Ave., to facilitate the storage and transfer of merchandise, such as grocery products, hardware, and other dry goods. The AT&SF accommodated these businesses by building several spur lines right up to their loading docks. Another major user of these spur lines was the Hahn Coal and Wood Yard, located east of the tracks and north of Railroad Ave. Wholesale warehouses and lumber yards, including the first office of a long-time Albuquerque business, the J. C. Baldridge Lumber Company were located south of Railroad Ave. along 1st St.

By 1900, New Town Albuquerque (now a fully incorporated city of more than 6,000 people) supported a wide range of small manufacturing and service industries including: brickyards, tanneries, flour mills, packing houses, wagon factories, steam laundries, bottling works, ice companies, and a cement plant.[9]

"Soon after its founding, the new townsite was divided administratively into four quadrants or wards. The wards, labeled First, Second, Third and Fourth (starting from the northeast quadrant and moving clockwise) were formed by the intersection of Railroad Ave . . . and the

Figure 4.1. Charles Ilfeld Company Warehouse in Albuquerque adjacent to AT&SF tracks, built 1911–1912, as it was being demolished in 1977. Photo by Jerry Goffe; Historic American Buildings Survey.

railroad tracks." The axes of this schema were determined by the rail yards and in particular by the Repair Shops, which were by far the young city's largest employer and the key to its development. "Even as the depot was being finished, the AT&SF shops and maintenance yards were under construction. By the mid-1880s, the locomotive and car-repair shops, and the roundhouse were completed. Within twenty years, 52,000 freight cars were passing through the city annually, and its shops and passenger facilities represented an investment by the company of more than \$3.5 million . . . [by 1893], a public elementary school was opened within each of the city's four wards."[10]

"To the south of downtown in the Third Ward, the Atlantic &
Pacific Addition and the Baca Addition together with smaller platted
additions that sometimes only encompassed a block or two were platted in the 1880s. The housing developments primarily served the
AT&SF employees who worked at the nearby locomotive shops and rail
yard. The housing stock was comprised primarily of small sturdy wood
frame cottages."[11] Without doubt New Albuquerque was a railroad
town, and it resembled other railroad towns throughout the West.
There was one respect, though, in which it was different: its size. By the
end of the 1890s, it was clearly destined to be the largest community
on the Santa Fe main line between Kansas City and Los Angeles thanks
in large part to the presence of the Locomotive Repair Shops.

CHAPTER 5

Overhauling Steam Locomotives

B y today's standards, steam locomotives of the final two decades of the nineteenth century were temperamental machines. As Dodge writes, they "required substantial daily servicing and maintenance, as well as periodic major overhauls. At the end of a daily run, a steam engine had its ash pan dumped and its firebox cleaned, ridding it of clinkers, the irregular lumps of coal left after firing. Its appliances and running gear were inspected, and, if necessary, repaired. Tubes, flues, and smokeboxes were cleaned. Boilers were washed out to remove mineral build-up approximately once per month, more frequently as necessary."[1]

In the early 1880s, when the AT&SF Locomotive Repair Shops went into operation at Albuquerque, the daily run of a steam locomotive was a roundtrip ranging from 200 to 300 miles, at the mid-point of which lubrication, inspection, and dumping of the ash pan were performed before the second leg of the roundtrip.[2] This daily cleaning and maintenance took place at roundhouses located at periodic points along the main track, usually 100 to 150 miles apart. These locations were known as "division points." The Albuquerque shops were at a division point, as well as being the site of shops that performed major overhauls, rebuilt locomotives, and repaired other rolling stock. The neighboring division points were at "Las Vegas to the north, Gallup to the west, and San Marcial to the south."[3]

Beyond this daily maintenance, the railroad specified in detail a series of monthly, quarterly, semi-annual, and annual inspections that had to be performed on each and every locomotive. By 1934, the standard monthly inspection included twenty-five individual items, ranging from examining and testing electric headlights to tightening unions and clamps on all sand pipes and flushing out tender cisterns.

Also mandatory were periodic, complete disassembly and rebuilding of every steam locomotive, what were called "heavy repairs," or "overhauls."[4] "Locomotives were taken to large shops, such as the ones at Albuquerque, for a major overhaul. In the nineteenth century, this might be necessary after as few as 40,000 miles."[5] Over the decades during which the Shops operated, the procedure involved in performing an overhaul remained very much the same, although the process became increasingly complex and mechanized over time.

That mechanization is especially evident in the use of very large traveling cranes that were capable of hoisting and moving entire locomotives. Cranes of this type were integrated into the "new" shop buildings erected at Albuquerque in the 1910s and 1920s (see chapter 11). Those cranes and the use of individual electric motors to operate power equipment of colossal scale were some of the most obvious changes between the original 1880s shops and their twentieth-century counterparts. Major locomotive repair shops, like those at Albuquerque, "required cranes, heavy tools, large stores, and ever-increasing manpower, as did the car repair shops that sprang up along the Santa Fe [line] to service the growing fleet of freight and passenger equipment."[6] The Albuquerque Shops encompassed both locomotive and car repair facilities.

Furthermore, the size and complexity of locomotives increased steadily over time. That meant that the number of individual steps that comprised an overhaul grew and grew and, with that, the technical know-how required of the shopmen who carried out the work. The physical size of the Shops, too, increased as they were enlarged to accommodate ever-larger rolling stock, as well as a swelling volume of railroad traffic. There were many other, though less apparent, changes, which we discuss in chapter 11.

For now, though, we outline the overhaul procedure, which in general would have been familiar to shopmen any time from the 1880s to the 1950s. A locomotive scheduled for periodic overhaul traveled under

Table 5.1. Required Inspections on AT&SF Steam Locomotives

Monthly Inspection (25 items)

1. Examine, grind in boiler and line checks, clean opening in boiler.
2. Grind, clean, and repack waterglass and gauge cocks and drain valves.
3. Remove caps from tank hose strainer box. Examine strainer and threads.
4. Wash out tender cistern.
5. Examine tank valves and strainer and see that valves are properly secured to stem.
6. Examine plates and brasses in tender.
7. Clean drain pipe back of tank.
8. Operate 3-way valve to reversing cylinder, and apply seal.
9. Examine and clean cylinder and channel cocks.
10. Examine and test electric headlight.
11. Examine blow-out pipes and force oil to engine and tender truck center casting.
12. See that radial buffer oil and grease holes are open. Close and pack oil caps.
13. Hayden exhaust pipe. Clean carter if necessary and record size of opening.
14. Test superheater units if necessary.
15. Inspect and repair stoker.
16. Wash out feedwater heater.
17. Test out feedwater heater for leaks.
18. Test out Klease feedwater pump.
19. Test out Klease feedwater motor.
20. Test air hose on locomotive and tender with soap suds.
21. Check grate shaker post and bar with template.
22. Test out mechanical lubricator to engine truck.
23. Test out mechanical lubricator to cylinders and valves.
24. Oil outboard bearings of multiple throttle.
25. Tighten unions and clamps on all sand pipes and line up with rail.

Quarterly Inspection–Add to Monthly (25 items)

1. Remove and examine engine truck brasses (passenger engines).
2. Remove and examine trailer truck brasses (passenger engines).
3. Remove and examine tender truck brasses (passenger engines).
4. Examine cylinder packing (freight and passenger engines).
5. Examine main valves, packing & valve stem fits on all road engines.
6. Examine and clean branch pipes if needed.
7. Remove and inspect drawbar, safety bars, safety chain, and pins.
8. Have items on 1154 card given inspection and tested.
9. Wash and inspect air pumps.
10. Clean air pump oil cup to air end.
11. Lubricate power reverse gear cylinder with brake cylinder lubricant.

Table 5.1 (*continued*)

12. Inspect stoker trough.
13. Inspect equipment in tender toolboxes.
14. Clean steam gauge siphon pipe and openings in connections into boilers.
15. Examine check valve feedwater heater and [illegible] line.
16. Clean out flange lubricator.
17. Are flange oil piping and nipples open?
18. Blow out hydrostatic lubricator to cylinders and valves.
19. Examine and clean stems in lubricator and oil pipes.
20. Clean and test out mechanical lubricator to engine trucks.
21. Clean and test out mechanical lubricator to cylinders and valves.
22. Clean air pump strainer.
23. Remove flange oilers. Dismantle, inspect, and clean all parts.
24. Remove and inspect valve gear parts, apply Magnaflux or white wash test and record test on plate attached to valve gear frame.
25. Check contours of radial buffers engine and tender.

Semi-annual Inspection–Add to Monthly and Quarterly (8 items)

1. Remove and examine engine truck brasses (freight engines).
2. Remove and examine trailer truck brasses (freight engines).
3. Remove and examine tender truck brasses (freight engines).
4. Examine cylinder packing (switch engines).
5. Examining main valves, packing and valve stem fits (switch engines).
6. Anneal, inspect, and stencil drawbars and pins.
7. Wash out 9 ½" and 11" pumps.
8. Clean and inspect mechanical lubricator to engine truck.

Annual Inspection–Add to Monthly, Quarterly, and Semi-annual (5 items)

1. Remove, clean, anneal, and inspect copper pipe to water column and left water glass cock.
2. Remove and clean water column.
3. Clean with rose-bit all fittings to boiler for water column and left water glass.
4. Remove dome cap and inspect throttle valve and rigging.
5. Remove receiving and discharge valves from 6" or 8½" cross stop and air compressors, which are to be thoroughly cleaned and inspected for defects. Parts found defective should not be continued in service.

Source: AT&SF Locomotive Inspection and Repair Report, Form 1215-A, Locomotive Folio, June 1, 1934, held by the New Mexico Steam Locomotive & Railroad Historical Society, Albuquerque, NM.

its own steam or was towed to the shops and was either pushed or driven on a spur track into a bay in the spacious machine shop. Within each bay was a long, masonry-lined pit that the locomotive straddled so that its undercarriage could be inspected and worked on. At each bay, there were usually "six to eight [large] machine tools (usually lathes and mills). These machines were used to do fine cutting required to fit engine parts."[7] Within the machine shop, fixed at specific locations, were other large machine tools, including multiple engine lathes of varying sizes, centering machines, bolt lathes, radial and vertical drills, bench drills, driving wheel lathes, axle lathes, journal lathes, quartering machines, at least one wheel press, metal saws, external and internal grinders of different diameters, bushing presses, assorted turret lathes, a nut facer, bolt cutters, a pipe threader, slotters, grooving machines, planers, crank shapers, graduated boring mills, drop tables, multiple arc welders, jib cranes, and on and on.[8]

Figure 5.1. AT&SF locomotive #3914 being dismantled at the Albuquerque Shops in February 1948. Photographer unknown. Courtesy of the Albuquerque Museum, Photo Archives, catalog number PA1980.184.896.

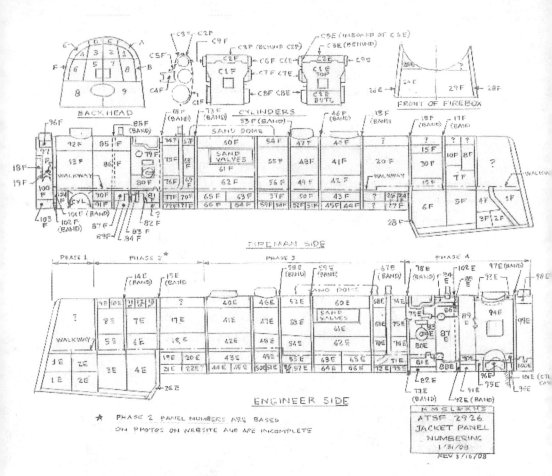

Figure 5.2. Jacketing plan for AT&SF locomotive #2926, prepared in 2008 during its restoration. Courtesy of the New Mexico Steam Locomotive & Railroad Historical Society.

After the locomotive had cooled down, its firebox and ash pan had been emptied, and its boiler drained, a crew of disassemblers swarmed over the body of the locomotive, removing every appendage, from lubrication tubes to the sand and steam domes on top of the barrel of the locomotive, as well as levers, lights, the smokestack, valves, gauges, cylinders, pistons, and the entire power train.

With exterior lines, pipes, and pieces removed, it was time to "strip" off the sheet metal jacket that sheathed the locomotive body. That jacket was composed of about 200 separate pieces of sheet metal that had been custom cut and shaped to fit around the various ports, channels, tubes, pipes, and conduits that passed through it. Those sheet-metal puzzle pieces, welded together to form the enveloping jacket, were virtually impossible to remove without damaging them. Thus, they would all have to be refabricated in order to rejacket the overhauled locomotive.

Beneath the jacket lay insulation, called lagging, the purpose of which was to reduce heat loss from the boiler that would render the engine less efficient and would subject the engineer and fireman working in the cab to intolerable temperatures. Although in the early days of steam railroading limited insulation was provided by closely spaced wooden slats strapped around the boiler, by the 1880s asbestos batting was the near universal material of choice. Removal of such insulation from the exterior surfaces of boilers, generating clouds of asbestos fibers, presented one of the many health hazards to which shopmen were routinely exposed.

After the insulation had been stripped, the boiler-firebox unit was removed for cleaning and restoration. That entailed replacing the dozens of pipes, or flues, that generated and then conducted steam to the cylinders and pistons that drove the locomotive's wheels. The cab would be removed and finally the locomotive body would be disconnected and lifted free of the running gear (drive wheels, trucks, and bogeys; see appendix 3: Steam Locomotive Components).

The next step was cleaning all the oil- and grease-coated, coal-dust-encased parts. The smaller parts were dumped in lots into a hot lye bath. The larger parts, such as drive rods and wheels, were power steam cleaned. After they'd been rinsed and dried, all parts were inspected for damage and checked against size and mating specifications. Engineering drawings prepared at the time of manufacture of the particular locomotive recorded those specifications, which were subject to minor alterations noted in repair and maintenance logs.

Those parts too damaged to permit repair had to be replaced by newly fabricated duplicates made in the blacksmith shop, sheet metal shop, or babbit (bushing) shop. Because of this, "the blacksmith shop, which supplied both the machine and freight car shops with forgings, occupied an increasingly critical position."[9] That was because there was almost no standardization of steam locomotive parts; each new locomotive model, usually manufactured in runs of no more than a dozen units, was customarily redesigned from the running gear up. As Albert Churella aptly writes, "Steam locomotives remained customized, purpose-built machines, and the necessity of tailoring locomotive designs to specific railroad requirements ensured that economies of scale were largely unobtainable" in the manufacture of parts.[10]

By the first decade of the 1900s, blacksmith shops were becoming increasingly mechanized. New machines included "oil furnaces, blast equipment, presses, steam hammers, and powered threading and shearing machines. As well as the manufacture of forgings, the heat treatment of metal ensured that materials were hardened, tempered, and annealed."[11]

In the case of replacement locomotive cabs, sheet metal workers flattened out the components to be replaced and then traced the resulting patterns, from which brand new substitutes could be cut, reassembled, and fastened in place either with rivets or by welding.[12] Sometimes, serious damage to a cab required a complete redesign based on measurements taken from still intact portions of the locomotive. Then the new cab was reattached to the chassis.

Strict standards governed the fit of all moving parts. Excerpts from the twelfth edition of W. P. James's *Enginemen's Manual*, dated 1917, provide some examples. For locomotives destined for road service (i.e., not within a rail yard): "The bore of main rod bearing shall not exceed pin diameters more than three thirty-seconds inch at front or back end. The total lost motion at both ends shall not exceed five thirty-seconds inch." Furthermore, "Crossheads shall be maintained in a safe and suitable condition for service, with not more than one-fourth inch vertical or five-sixteenths inch lateral play between crossheads and guides." Crossheads are jointed rods that convert the reciprocal motion of the pistons into the rotary motion of the wheels.[13]

Among the myriad components of steam locomotives that had to be inspected and refurbished or replaced during a general overhaul were the large drive wheels. Out-of-round wheels or wheels with worn flanges—the lips that held the wheels between the tracks—were usually

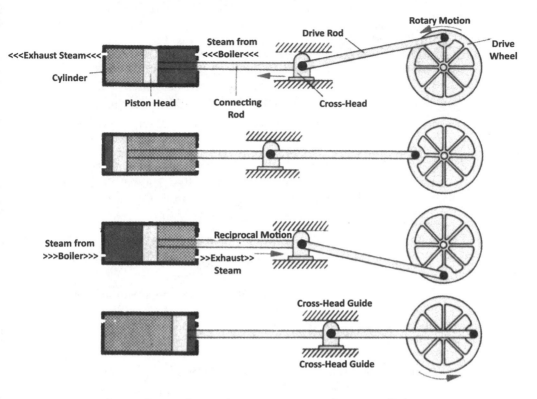

Figure 5.3. Schematic drawing of a steam locomotive's power train from steam cylinder to connecting rod, through the cross-head and cross-head guide, to drive rod and drive wheel. This articulation of parts converts reciprocal motion (in the cylinder) into rotary motion (of the wheel). Author's drawing, adapted from *The Way Things Work, An Illustrated Encyclopedia of Technology*, Vol. 1 (New York: Simon and Schuster, 1967), 43.

repaired by replacing the "tires," the outer, steel rims. These steel tires were mounted and dismounted by heating and expanding them and then cooling and shrinking them onto the wheels. In addition, wheels of different diameter could be installed for special purposes. Generally, freight locomotives ran with smaller diameter wheels than did passenger locomotives.[14]

With thousands of parts repaired or replaced, it was time to reassemble the refurbished jigsaw puzzle. Again, there was little margin for error. Torqueing of bolts and nuts was carefully gauged. The cutting and cementing of gaskets were precision jobs. The brightness of headlights was a matter of clear visibility of a dark object the size of a man at the distance of 1,000 feet. To ensure wheel traction in wet or icy conditions, "sand pipes must be securely fastened in line with the rails."[15] The *Enginemen's Manual* specifies 158 different conditions of a steam locomotive—like these regarding headlights and sand pipes— that had to be met before an engine could be returned to service. Then the locomotive had to be painted (in the United States usually uniformly black in the nineteenth and early twentieth centuries). And finally it had to be fired up and run under road conditions to make sure all those parts meshed and ran smoothly, without catches or slippage, without the least wobble or excessive friction, and without leakage of steam.

At the Albuquerque Shops, up to thirty locomotives were in various stages of the overhaul process on any given day. At the Shops' peak, crews were able to fully overhaul an average of forty locomotives a month, month in and month out, despite injuries to and absence of workers and the occasional intractable mechanical problems of locomotives.

A seventeen-minute film from 1938 of steam locomotive overhaul work at the London, Midland & Scottish (LMS) Railway shops in Britain includes brief glimpses of the same general procedures as were followed at the Albuquerque Locomotive Repair Shops described above.[16] There was one major difference, though, between the procedures followed at the LMS shops and those at the AT&SF's Albuquerque Locomotive Repair Shops. The British system ran on an assembly-line or longitudinal model in which the locomotives were arrayed for overhaul in single file on a moving conveyer. The defect of that system was that if there was a problem with disassembly or reassembly of even just one locomotive, the whole line came to a standstill. At Albuquerque, on the other hand, each locomotive was stripped down and reassembled in a stall by itself. There were twenty-six parallel, side-by-side stalls, in what was called a transverse layout. Thus, if one or more engines presented complex problems, their extended overhaul time did not interfere with work on other locomotives in their own separate bays.[17] That difference aside, the LMS video offers people of today an opportunity to observe the same sort of dismantling and rebuilding work as was the daily routine at the Albuquerque Locomotive Repair Shops.

Figure 5.4. Photo of locomotives under repair in the erecting bay of the machine shop at the Albuquerque Shops, February 1948. View to southwest. Note the ends of concrete-lined inspection pits to the right of each locomotive, the scrap car in the foreground, an overhead traveling crane in background, and the 256-ton traveling crane at the top foreground. Photographer unknown. Courtesy of the Albuquerque Museum, Photo Archives, Catalog No. PA1980.184.895.

The work of locomotive overhaul was a coordinated undertaking involving shop employees from the various departments: inspection, machine shop, boiler shop, blacksmith shop, sheet metal shop, babbitt shop, paint shop, and store, as well as assorted smaller units such as the stationary engineer, electric plant, and parts vat. There were six major craft specialties represented among railroad shopmen: machinists, boilermakers, blacksmiths, car repairers, electricians, and sheet metal workers, plus many other more minor, but necessary, shop specialties, including painters, carpenters, and draftsmen.[18]

Teams, or gangs, of shopmen worked together on individual engines, from initial stripping to reassembly-erection. Experienced teams worked smoothly together, following a set order of coordinated work. The resulting interdependence of shopmen made for a high degree of camaraderie and trust among the shop workforce. The shopmen all faced the same challenges and dangers every day, which required them to work together with confidence in the skill and care of their work-mates. And each had a vested interest in the proficiency of the other members of the team, which at its best worked as smoothly as a well-serviced locomotive. As with any professionals with journeymen's level of experience, the steps required by locomotive overhaul were second nature to every member of a practiced team. Unspoken work habits and innovative techniques were an integral part of work at the Locomotive Repair Shops. But such minute details of shop work were rarely recorded and have largely been lost to memory.

Often, specific operations required the skill and muscle of several men. For example, removing the heavy steel, main drive rod from the crank on the main drive wheel on one end and from the crosshead pin on the other end required the care and strength of five or six men. Although cranes, lifts, and dollies were used to move rods weighing 800 to 1,500 pounds each around the shops, there was, in the end, no substitute for coordinated application of human muscle power and guidance by human eyes and hands in the precision removal or replace-ment of such heavy, cumbersome parts.

Forging large replacement parts, such as a drive rod, likewise entailed the work of a crew of three or four people to manage the mechanized, industrial-scale drop hammer and the white-hot steel blank. Besides a journeyman blacksmith, such a crew would include a couple of blacksmith helpers as well as an apprentice. Because of the constant, potential risk of injury associated with most tasks in the Locomotive Repair Shops, members of gangs—whether permanent or ad hoc—had to have a high degree of confidence and trust in their fel-low workers. That confidence and trust was called on throughout the shops, hour after hour, day in and day out.

In addition to the overhauling of locomotives at the Repair Shops, other gangs were responsible for refurbishing or complete rebuilding of a large array of different types of freight cars, from wooden boxcars to

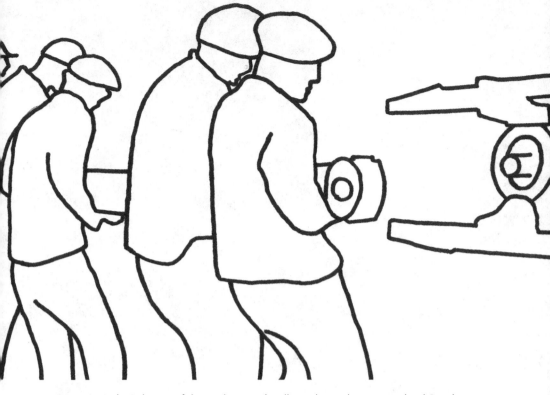

Figure 5.5. Author's drawing of shopworkers man-handling a drive rod onto a cross-head. Based on a frame from *General Repair*, a short 1938 film documenting the complete rebuild of a steam loco-motive at the shops of the London, Midland & Scottish Railway.

gondola and hopper cars. The accompanying illustration shows engi-neering specifications for a truck (four-small-wheel assembly) with a forty-ton load capacity. This type of truck was in use on freight cars during the early 1900s and would have been familiar to Albuquerque shopworkers.

No matter how many hours or how many days shopmen worked, they were responsible for the hand tools they used, not the normal wear and tear, but their loss or disappearance. When a machinist, a boiler-maker, a blacksmith, a sheet metal worker, or an electrician needed a specific tool, he went to the large caged tool room that occupied the northern middle of the heavy machinery bay and requested the tool from the tool supervisor. When the tool was then delivered to the shop-man, he had to sign and date a loan form, which he would later have

Figure 5.6. Hammering out a drawbar under the steam hammer at the Atchison, Topeka and Santa Fe Railroad blacksmith shop, Albuquerque, NM. Photo by Jack Delano, 1943.

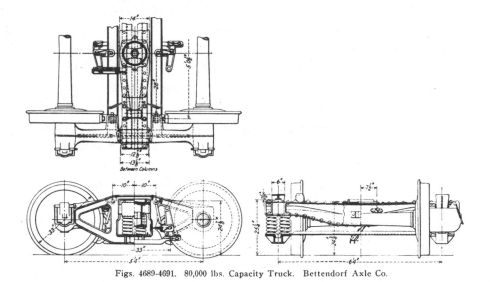

Figs. 4689-4691.   80,000 lbs. Capacity Truck.   Bettendorf Axle Co.

Figure 5.7. Engineering drawings of an 80,000-lb-capacity truck (four small wheels connected by a frame in tandem). From Francis E. Lister, comp., *Car Builders Dictionary*. (New York: Railway Age, 1909), 477.

Figure 5.8. Author's photo of the tool room in the machine shop at the Albuquerque Locomotive Repair Shops as it looked in 2017.

to sign and date again when he returned the tool to the supervisor. The shopman was liable for the replacement value of the tool, if he was not able to account for its breakage or loss and was not able to return the tool. For the entire time he needed the tool, and that might be years, he could store it at the end of each day in his own locked toolbox, which was kept near or at his current workstation. But eventually, when he left the employ of the Shops or retired, all those loan forms had to be redeemed or cancelled. So the borrowing and returning of tools was another rhythm within the overall routines of work in the Shops.

Whether at the lye bath or the sheet metal shop, the boiler shop or the roundhouse, the store or the babbitt shop, the light machine bay or the overhead cranes, specialists throughout the shop complex engaged every hour of every workday in the myriad tasks involved in keeping AT&SF rolling stock running with a minimum of risk to passengers, goods, and other railroad personnel. Make no mistake, though, most jobs at the Shops were dangerous and regularly resulted in injury to shopmen.

Because faulty or damaged boilers were far and away the leading cause of catastrophic steam locomotive failures, the repair and refabrication of combination boiler-fireboxes were crucial steps in the overhaul process. A machinists' journal called *The Locomotive* reported for many years quarterly figures on locomotive boiler explosions in North America and their fatal outcomes. The April 1904 issue is typical, reporting forty-five locomotive boiler explosions in the United States and Canada during the preceding year, including three on the AT&SF. One of those Santa Fe explosions was recorded in this way: "The boiler of locomotive No. 471 of the Atchison, Topeka & Santa Fe Railroad exploded, on April 13th, at Florence, Kan. Fireman Hauhn was killed, and engineer Moody was injured so badly that he died shortly afterwards. The boiler of the locomotive was completely demolished."[19]

To minimize such tragic disasters, boiler designs were continually tweaked and improved, and national standards for maintenance and repair of boilers were frequently strengthened. Also, throughout the steam era, work for boilermakers and boiler smiths was always available. And it was always dangerous. For one thing, boiler inspections were carried out under pressure, which meant that being engulfed by or sprayed with live steam was a daily and deadly possibility for workers in boiler shops.

Many were the dangers to machinists, boilermakers, and their apprentices and helpers during the process of disassembly of locomotives and the repair of parts. Steam locomotives were aggregates of 6,000–7,000 metal pieces and parts held together by bolts, rivets, and welds, which over months of nearly constant vibration and impact could jam, break, crack, or simply fall off.[20] Disassembly therefore routinely required that manual force be applied to bent, twisted, or otherwise distorted connectors. Oversized wrenches wedged onto the heads of frozen bolts and then beaten with sledgehammers were a

Figure 5.9. Photo of C&O steam locomotive #3020 after the explosion of its boiler, May 1943, at Chillicothe, OH. As was typical of such events, the explosion extruded the boiler flues through the front end of the boiler, creating a spaghetti-like tangle. Photographer unknown. Courtesy of C&O Historical Society, item number COHS-6198.

continual recipe for mashed and broken fingers, as well as flying parts of broken tools. Manhandling heavy, grease-coated parts such as drive rods put human backs and muscles at risk every day. Slippery, oily surfaces, paired with equally slick soles of shoes made for treacherous footing around and on locomotives being overhauled. Sharp edges of sheet metal, metal shavings, and splinters as well as hot welds and solder were constant hazards. All of this was compounded by the system of constantly moving belts that drove larger machine tools and by rotating gear-driven machines that could snare hands, arms, feet, legs, and loose clothing. And the situation was aggravated by the nearly incessant clanging and banging of metal on metal. In 1885, a visitor to the Topeka, Kansas, shops of the AT&SF "reported that many years of work in the boiler shops were literally deafening."[21]

According to Mike Baca, even as late as 1945, many of the larger machines in the Albuquerque Shops ran off long belts powered centrally by a steam engine. That arrangement ramped up the noise in the Shops even more.[22]

In the face of such a multiplicity of hazards, nineteenth- and early twentieth-century shopworkers routinely had no safety equipment available beyond gloves and welder's goggles. Even so, many workers managed to avoid serious injury while working at the Shops, but virtually no one escaped the daily cuts, scrapes, bruises, burns, contusions, and eye injuries that went with the job of overhauling steam locomotives. As but one example, in the early 1940s at the Albuquerque Shops, pipe fitter apprentice Tony Gutiérrez's clothing got caught in the rotation of a pipe threading machine. He was saved from serious injury only by quick action by his brother at the next bench, who cut power to the threader and stopped it.[23]

Everybody, shopmen and their families and railroad management, knew that work at any locomotive repair shop or at any roundhouse was inherently dangerous.[24] Issues of the *Santa Fe Magazine* for employees frequently carried reports of on-the-job injuries to shopmen. The September 1911 issue, for instance, breezily reported that "C. L. Berndtson and wife have returned [to Albuquerque] from Denver, where they spent several weeks. Mr. Berndtson had the misfortune to severely mash his foot and thought it a pretty good time to take a vacation."[25] Somewhat more cryptically, in December 1914, the magazine noted that "F. Salazar, machinist apprentice [at Albuquerque], was injured in a runaway. He is off duty." Further, "[Albuquerque] Boilermaker E. A. Moon has been off duty on account of an injury to one of his eyes, caused by a piece of a chisel breaking the glass in his goggles. A piece of the glass flew into his eye."[26] In 1940, the hazards of shop work were still ever present; machinist Ray Stewart suffered a double hernia.[27]

To keep trained, experienced shopmen on the job was naturally an important goal of AT&SF management. Starting over with brand new apprentices meant four years of less than optimal work from each understudy. Training new journeymen was a costly undertaking. So,

Figure 5.10. Santa Fe Railway Hospital at Albuquerque, 1880s. Cobb Studios photo. Courtesy of Center for Southwest Research, University Libraries, University of New Mexico; Cobb Memorial Photography Collection 000-119-0799.

when shopmen suffered serious injuries on the job, the Railway sought to get them the best and quickest curative care available.

> At least by the 1880s the Santa Fe had company-appointed doctors in many towns on the line. In an attempt to deal more systematically with the problem of injured and ill workers, Santa Fe Vice-President A. E. Touzalin on March 1, 1884, issued a circular announcing the inception of the Atchison Railroad Employees' Association. . . . The purpose of the organization was to build hospitals and contract with doctors throughout the line to care for sick and injured employees. . . . Indeed, the A&P had established a similar organization and built a hospital in Albuquerque in 1882.[28]

In cases in which an injured employee sued the Railroad, alleging its responsibility for unsafe conditions or reckless supervisors, management often settled at reasonable terms. Not only did that tend to satisfy the injured employee, but also it helped to assure other shopmen that they, too, would be treated fairly, if they were unlucky enough to suffer a serious injury. In sum, the Santa Fe often "treat[ed] those involved in accidents more liberally than was legally necessary." That, in turn, encouraged company loyalty. On a related matter, "the Santa Fe had no pension plan until 1907. However, old-timers too enfeebled to carry on their usual jobs were sometimes given easy tasks they could handle," and thus were able to maintain employment.[29]

CHAPTER 6

Job Specialties

From its opening in 1881, the Albuquerque Locomotive Repair Shops had constant need of a large, skilled workforce augmented by less skilled laborers, helpers, and apprentices. Each facet of locomotive overhaul or repair required the work of teams of trade specialists, each skilled in and intimately familiar with the procedures for disassembly, reassembly, cleaning, fabrication, painting, and lubrication of particular sets of components that made up the many locomotive models. Each sub-assemblage of the locomotive mechanism was the responsibility of titled specialists, and each specialty had its own professional ladder.

An exhaustive list of job categories within a typical steam locomotive repair shop of a large American railroad between 1880 and 1955 is impossible to reconstruct because the workforce of each shop had to include a mix of workers who matched the assortment of locomotive models in use by that particular railroad at the time. And the stable of models was constantly in flux. As Albert Churella correctly observes, "Even slight variations in operational requirements could require steam locomotive designs to be altered substantially. As a result, steam locomotive designs proliferated."[1] Each issue of the trade journal *The Railway Mechanical Engineer*—published monthly from 1916 through 1949—was filled with scaled drawings of design improvements and new apparatus for steam locomotives. Not all of these ideas were realized in parts and assemblages that came into wide use, but literally thousands of them did.

Within the welter of different designs, there were classes or variations on an original innovation. There were many such classes that came out of the mechanical engineering department of the Santa Fe Railway in Topeka, Kansas. Larry Brasher's abundantly illustrated chronology of AT&SF locomotive development covers scores of locomotive classes. To single out just three general designs from the AT&SF, "The 3460 class, the 3765 class, and the 5001 class have rightly been called the Santa Fe's 'Big Three.'"[2]

To give a general idea of the number of different job specialties that needed to be represented among the repair shop staff, we provide the following list of job titles reported by AT&SF shopmen at Albuquerque in 1896, as reflected in the city directory for that year. The list of sixty-one distinct job categories, by no means an exhaustive catalog of all the job titles of Albuquerque shopmen, hints at the variety of work done by them. Hidden within classifications such as "machinist," though, were subspecialties including special skill with boiler flues, superheaters, air-brakes, side rods and wheels, valves and gauges, steam pistons and cylinders, control linkages, sand pipes and water supply lines, and automated stoking devices, to name just a few. As stated in a 1923 description of the Santa Fe apprentice program, when employees completed the training period, "they should be assigned to some particular class of work for which they have shown a peculiar adaptability and upon which they will soon become experts."[3]

All Shop departments required a core staff of journeymen, that is, workers who had gained the requisite training and experience to be accredited as fully capable practitioners of their trades. AT&SF required that journeymen machinists, boilermakers, blacksmiths, and other skilled workers complete a four-year apprenticeship. During that time of supervised labor, combined with formalized instruction, each apprentice was required to learn to perform competently the routine tasks associated with his specialty. Furthermore, the apprentice had to receive instruction that would help in dealing with complex problems that were bound to develop during the course of regular work. Problem solving was a skill every journeyman had to demonstrate. At the shops there was no such thing as a routine overhaul; invariably there were

Table 6.1. Employee Titles, Albuquerque Locomotive Repair Shops, 1896

asst. car clerk
asst. foreman, car dept.
blacksmith
blacksmith apprentice
blacksmith helper
boilermaker
boilermaker apprentice
boilermaker helper
caller at the roundhouse
car cleaner
car clerk
car inspector
car laborer
carpenter
carpenter foreman
car repairer
car repair laborer
chief bill clerk
chief clerk
chief clerk to gen. master mechanic
chief clerk to storekeeper
clerk
clerk to the car accountant
clerk to storekeeper
daytime clerk
employee in the storehouse
foreman
foreman at the roundhouse
foreman in the car dept.
foreman in the tinshop
foreman of laborers
foreman of the boiler shop
general master mechanic
general superintendent
helper
hostler at the roundhouse
laborer
laborer for storekeeper
machinist
machinist apprentice
machinist helper

messenger
night watchman
operator
painter
pattern maker
seal taker
stationary engineer
stenographer to master mechanic
storekeeper
tank repairer
tinner
tinner apprentice
tinner helper
upholsterer
washer at the roundhouse
watchman
watchman at the storehouse
wiper
yard foreman

Total of 61 job titles

Machinists: 39
Machinist apprentices: 12
Machinist helpers: 2

Boilermakers: 9
Boilermaker apprentices: 5
Boilermaker helpers: 19

Blacksmiths: 12
Blacksmith apprentices: 2
Blacksmith helpers: 17

Source: Roberts, *Albuquerque City Directory and Business Guide for 1896.*

complications that made every locomotive unique, requiring imaginative solutions to apparent impasses. An article in *The Railway Age Monthly and Railway Service Magazine* for May 1880 put it this way: "While the take-it-easy mechanic, whose leading ambition is to put in a certain number of hours a day and get away from the shop, is bothering the foreman for instructions in overcoming some difficulty, his thinking fellow worker contrives a plan of his own and accomplishes the desired object. The demand is for more mechanics who think, not only in the shop but out of it—those who probe outside sources of information in order to advance themselves in those qualifications which are sure to command recognition."[4]

Apprentice training was not exclusive to the AT&SF, nor was it new in the world of work. Originating in the distant, undocumented past with the transmission of know-how among close relatives, formalized systems of apprenticeship had existed for artisans and tradespeople in much of the world since at least the late Middle Ages.[5] The immediate antecedents of the Santa Fe Railway's apprentice program were European industrial apprenticeships of the nineteenth century, and the company's version of locomotive repair apprenticeship echoed those European models in many respects.

Although the AT&SF had an apprenticeship program from at least the 1910s, it became even more important as the 1922 shopmen's strike took hold and dragged on for months and into a second year. As one management observer spelled it out, during the "crisis . . . we have had to take into the shops inexperienced men to carry on the work that was formerly carried on by mechanics with years of experience." Large numbers of striking or locked-out employees were replaced, and clearly their substitutes had to be quickly brought to an acceptable level of competence. E. H. Hall reported, "I had an opportunity a few months ago to visit the shops of the Santa Fe at San Bernardino, Cal. It was wonderful to see the work they were accomplishing with inexperienced men, who had never seen the inside of a coach before and had only been there 30 days."[6]

Without referring explicitly to the strike, a Santa Fe official touted the benefits of the AT&SF apprentice program, which not only produces "first-class mechanics," but "also cultivates a spirit of loyalty to

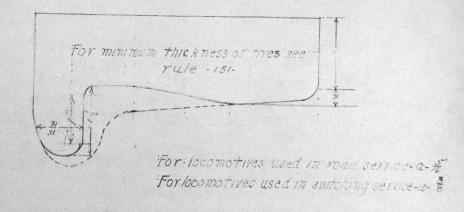

For minimum thickness of tires see rule -151-

For locomotives used in road service - a - $\frac{5}{16}$
For locomotives used in switching service - a - $\frac{3}{8}$

SANTA FE
TIRE DEFECT
AND LIMITS OF WEA
R.L. MORTON
JUNE- 3 - 25-N0-1

Figure 6.1. Drawing by AT&SF apprentice Al Morton showing railroad tire in cross section to delineate a defect and limit of acceptable wear. Courtesy of WHEELS Museum, "Apprentice School Drawings, Waynoka, OK.," June 23, 1925.

the company which money cannot buy."[7] Such results were likely because young apprentices were to be "boys" as young as sixteen (likely without union experience), who would often be taking the places of much older union members. At a minimum, "all boys entering an apprenticeship should be able to add, subtract, multiply, divide and work simple and decimal fractions," and their apprentice training would include being "taught to read a blueprint and make working sketches."[8] Former Albuquerque shopmen Mike Baca, Eloy Gutiérrez, and Thomas Cordova all remembered attending mechanical drawing and blueprint classes, as apprentices, in the evenings at the Albuquerque Shops' General Administration Building.[9] To assure that each apprentice could be given individual attention, "there should be a shop instructor appointed for every 25 or 30 boys in the shop."[10]

Later, during the Second World War, repair shop apprentices were eligible for deferral from military service. But in a perverse logic, once an apprentice completed his training, he could immediately be drafted.[11] One effect of this policy was that the railroads were training many military machinists but not getting the full benefit of work by the newly trained journeymen.

Idealized portrayals of the railroad apprenticeship system were frequently belied by social, ethnic, and racial prejudice, as well as budgetary concerns of railroad owners and managers. As historian Colin Davis writes, "Blacks were concentrated in the unskilled and semiskilled occupations, but in many cases African American 'helpers' and general laborers were, in fact, quite highly skilled and were kept out of higher classifications by virtue of racial restrictions rather than by their lack of talent or knowledge."[12] Davis's research concentrated on shopmen in the East and Midwest of the United States. If he had scrutinized as thoroughly railroad employment in the Southwest and West, he also would have found equally discriminatory job classification practices with respect to Hispanics, Native Americans, and Asians.

Several former shopmen at the Albuquerque Locomotive Repair Shops interviewed for this book reported chronic failure of skilled helpers and laborers of color to be advanced up the ladder of skill grades. Dr. Eloy Gutiérrez, well-known Albuquerque dentist and himself a sheet metal and pipefitter apprentice in the 1940s, told that his brother Frank was only the second Hispanic to become a foreman at the AT&SF Shops more than sixty years after the opening of the Shops. It should also be pointed out that Frank Gutiérrez eventually became equipment supervisor for the entire AT&SF system.[13] But for generations such advancement was barred to Hispanics, Native Americans, and Asians, as well as African Americans.

Likewise, Jim Brown writes in an undated manuscript, "Although there were opportunities for in-house training and job advancement [at the Albuquerque Shops], most of the skilled positions in the early 1900s were held by white men who had been trained elsewhere in the United States or in Europe."[14] As we have pointed out earlier, information from various annual editions of Albuquerque city directories support both Brown's findings and Gutiérrez's memory. By 1950, though, there was near parity between Hispanics (113) and non-Hispanics (120) among AT&SF journeymen machinists listed in the city directory for that year.[15] This is in keeping with the general Hispanization of the shop staff at Albuquerque beginning after the 1922 shopmen's strike (see chapter 13).

On the other hand, company management at all levels, from supervisors and foremen on up, was almost entirely in the control of non-Hispanic employees. Of the fifty-three Albuquerque Shops employees who were listed in the 1950 city directory as occupying supervisory positions—including superintendent, master mechanic, storekeeper, foreman, supervisor, and assistant foreman—only two were Spanish-surnamed. They were foremen and therefore at the lowest level of the supervisory staff.

That represents only the tiniest improvement from 1896, when all fifteen men in supervisory positions at the then much smaller Shops were non-Hispanic. Thus, throughout the steam era, while the rank-and-file staff of the Albuquerque Shops became increasingly Hispanic, the same was not true in the supervisory ranks.[16] Regardless of job title, there were pervasive ethnic differences in play throughout the Shops before WWII.[17] Age discrimination was also prevalent. Olivia Cordova Loomis remembered her father Thomas Cordova saying that when he was an apprentice, older men were often summarily let go, even if they were close to retirement age.[18]

The job classifications of shopmen also established their position within the company's professional ranking. The most obvious distinctions, indicated by pay level, were between skilled and unskilled workers. Thus, the skilled staff—electrical workers, sheet metal workers, machinists, boilermakers, blacksmiths, and carmen—were bunched at the top of the order, with average annual pay in 1950 ranging from just under $3,700 to about $3,850. Meanwhile, workers in the least skilled positions, laborers—with small variations among them—averaged in the neighborhood of $2,700 a year. A middle ground was taken up by apprentices and coach cleaners, at under $3,000 a year, and helpers to the skilled trades and helper apprentices who were somewhat better

Table 6.2. AT&SF, Number of Full-Time Employees and Amount of
Compensation, 1950

Job Classification	Total # AT&SF	Total Compensation	Avg. Annual Compensation*
Blacksmith	196	$718,453	$3,666
Boilermaker	514	$1,937,284	$3,769
Carman	3,697	$13,417,094	$3,629
Electrical Worker	592	$2,281,730	$3,854
Machinist	2,222	$8,378,744	$3,771
Sheet Metal Worker	541	$2,042,213	$3,775
Skilled Trades Helper	3,331	$10,747,977	$3,227
Helper Apprentice	75	$245,508	$3,273
Regular Apprentice	695	$2,034,601	$2,927
Coach Cleaner	704	$2,102,225	$2,986
Laborer	3,675	$9,932,207	$2,703

*Average US annual family income for 1950 was $3,319 according to the US Department of Commerce, Bureau of the Census. "Median Money Income of Families, 1:297.
Source: AT&SF Railway, *Annual reports of the Atchison, Topeka & Santa Fe Railway Company to the* [Kansas] *State Corporation Commission,* 1945–1959. Kansas State Historical Society, Item No. 218470, "Employees, Service, and Compensation, Chart 561, IV. Maintenance of Equipment and Stores," 522; average annual compensation computed by the authors.

paid, in the range of $3,200 to $3,300. Although these specific figures for pay apply only to 1950, the relative level of compensation among the various job classifications was typical of the entire seventy-plus-year lifespan of the Locomotive Repair Shops.

It is worth pointing out that systemwide within the Santa Fe Railway, shop staffs were divided fairly evenly between skilled workers (47.8% in 1950) and unskilled and semiskilled workers (52.2%), with the margin going to those with lower skill. It was, of course, in the company's financial interest to maximize the lower-skilled share of the shop workforce, thereby minimizing payroll. As we have already seen, employees regularly complained about company efforts to artificially skew those percentages by failing to promote workers even when their skill level would have warranted it.

CHAPTER 7

The Workforce

1880–1900

While many of the jobs involved in laying railroad track could be filled by individuals with minimal training, most of those dealing with the repair and maintenance of the rolling stock—the locomotives and cars of various types—required specialized skills and experience. By 1880, steam-powered trains had already been operating in the eastern United States for some fifty years but for only about a decade anywhere in the West. Therefore, initially most of the skilled positions at the Albuquerque Locomotive Repair Shops were filled by experienced railroaders from the East and Midwest and by European immigrants.

Accompanying the leading edge of the laying of track were not only road construction gangs, but also machinists, boilermakers, and other skilled journeymen. Without a knowledgeable crew with such technical know-how, trains simply could not be kept running. As we have already pointed out, locomotives of that period were complex, temperamental machines, subject to frequent malfunction and requiring daily inspection and periodic major repairs. So, it comes as no surprise that two days after the AT&SF's last rail was laid into Albuquerque, a crew boarding car and camp moved into town, bringing a contingent of mechanical specialists.[1]

Until a roundhouse could be built in Albuquerque, complex repairs on locomotives would have to be taken care of by a resident crew at the recently completed roundhouse in Las Vegas, New Mexico.

Erecting buildings that would comprise a locomotive repair facility was not by itself sufficient to launch the work of overhauling locomotives at Albuquerque. A permanent skilled workforce had to be hired and settled as residents. That turned out to be a more difficult task than AT&SF management imagined it would be, and securing and maintaining a staff would for years prove to be a challenge for the Railway. There was a paucity of native New Mexican machinists, boilermakers, and metal workers. At the same time, many Easterners with such skills considered the physical and social environments of New Mexico to be unappealing. As a result, they either chose not to accept employment at Albuquerque or did not stay long once they took jobs. Labor historian James Ducker comments, "In 1900 the *Railway Gazette* still considered the Southwest to be the hardest place in the nation to get quality workers. Recruitment of skilled and semiskilled employees in this region presented a problem the Santa Fe was never able to solve fully. Even at the end of the [nineteenth] century it found many of its workers in New Mexico and Arizona abandoning the line every summer to head for cooler climes."[2]

Albuquerque and New Mexico natives were unlikely to get technical jobs at the locomotive repair shops until some were able to complete the four-year apprenticeship required by AT&SF for an individual to achieve journeyman status. Even then, ethnic prejudice, which was endemic throughout the railroad industry, was a barrier to hiring and advancement of Hispanos, Native Americans, Asians, and African Americans. Workers in any of those groups found it difficult to be admitted to the ranks of apprentices.

Getting hired by AT&SF was complicated by the fact that the company preferred hiring family members of people who were already on the Santa Fe's payroll. As many researchers have remarked, knowing or being related to someone who already had a good employment record with the company always gave the prospective employee an advantage over those who did not. In part, this hiring practice of the AT&SF was a measure that led to a very loyal workforce. This kind of "favoritism encouraged [the employee] to remain loyal in the hope that he might be able to pass along his good fortune to his friends and relatives," and many did.[3]

Even a casual examination of the Albuquerque Shop payrolls or the

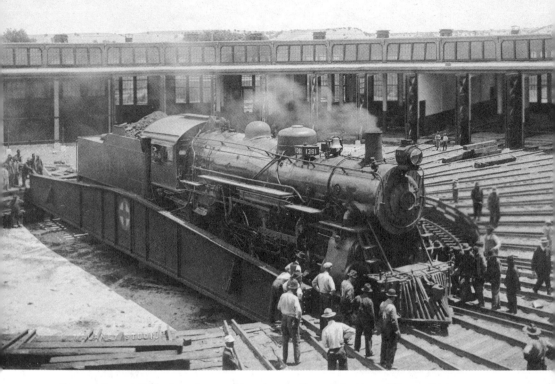

Figure 7.1. Locomotive #1361 on the turntable at the roundhouse in Las Vegas, NM, 1917. This was the first locomotive to enter the new roundhouse. Malaney Studio photograph. Courtesy of Center for Southwest Research, University Libraries, University of New Mexico; Gross, Kelly and Company Pictorial Collection 000-096-A10-0032.

City Directory shows unmistakably that many sets of close relatives were employed as shopmen at the same time. For example, the 1896 *City Directory* lists two men with the surname Bambrook working at the Shops, one as a machinist apprentice and the other as a pattern maker. Likewise, two men named Booth, living at the same address on South Edith Street were both employed as clerks at the Shops. And four men with the surname Britton were working at the Locomotive Repair Shops, three of them in the boiler shop. Consequently, newcomers to New Mexico being the earliest hires at the Albuquerque Shops meant that it would be years before locals could break into the ranks of shopmen in significant numbers, even ignoring other preferential factors. Even the rare Hispanics who worked at the Shops during the early years exhibited the same pattern of family hiring. Seven men with the surname Apodaca, for instance, all apparently living on Barelas Road, were simultaneously working at the Shops in 1896.[4]

Prejudice was nonetheless pervasive for Hispanics. "In 1880, native-born Hispanics comprised the largest group [of wage earners in Albuquerque], 234 persons, or 60 percent of all employed male heads of households. Over 80 percent of the Hispanics, however, were employed as unskilled laborers. Conversely, more than 40 percent of the native-born whites held skilled jobs, and almost one-third were employed in the proprietary/managerial field. In the third-largest group, foreign born, most male heads of households (over 40 percent) held skilled jobs, while proprietors, managers, and professionals provided over 34 percent of immigrant male heads of households [in Albuquerque]." Five years later "in all but the native-born Hispanic group, the majority of men were employed as skilled laborers."[5] The disparity between ethnic groups was starker within the locomotive repair shops, and it remained a stubborn fact even as an increasing number of Hispanics obtained jobs in skilled positions by 1900. It also appears that during the first twenty years of the Locomotive Repair Shop's existence "for Albuquerque's foreign-born population, there was a remarkable degree of economic success, both in occupational mobility and geographic persistence."[6]

The hurdle of ethnic bias, though, remained in place even when the company found it difficult to retain Eastern United States and European apprentices for their full four-year commitment. Machinist jobs with other, especially smaller, railroads were plentiful, as well as with other sorts of companies, such as the Albuquerque Foundry and Machine Works, located on the east side of the tracks opposite the Locomotive Repair Shops. Also, during the twentieth century, as the automobile increasingly became a factor in American life, the number of machinist jobs at car repair shops exploded. All of these job opportunities lured apprentices away from work for AT&SF.

By the 1890s, AT&SF adopted negative incentives in an attempt to hold on to apprentices for their full training tenure. "In 1892 the company began deducting 5 percent of the wages of the first year apprentices and 10 percent, 15 percent, and 20 percent of those in their second, third, and fourth years, respectively. At the end of the

Figure 7.2. Jacoby's Albuquerque Foundry, located on the east side of the AT&SF tracks opposite the Albuquerque Locomotive Repair Shops, 1880s. Courtesy of Center for Southwest Research, University Libraries, University of New Mexico; Cobb Memorial Photography Collection 000-119-0605.

fourth year [of apprenticeship] all of this was paid back to the apprentices."[7] Of course, if the apprentice left before the end of the fourth year, the deducted wages would be forfeited.

The relative overall scarcity of journeymen machinists, and the complete absence of Spanish-surnamed journeymen machinists, at the Albuquerque Locomotive Repair Shops during their first decades is confirmed by data extracted from the 1896 *Albuquerque City Directory and Business Guide*. The *Directory* lists thirty-eight machinists employed by the Locomotive Repair Shops, none of whom had Spanish surnames. Twelve machinist apprentices are listed, again none with Spanish surnames. The two machinist helpers, the next lower grade, once again included no one with an Spanish surname. Meanwhile, exactly half of the group of sixty-two laborers, the lowest level of employees who could repair and rebuild steam locomotives, had Spanish surnames.[8]

Likewise, the *Directory* for 1896 listed no Spanish-surnamed journeyman boilermakers at the Locomotive Repair Shops, no Spanish-surnamed boilermaker apprentices, and only 5 of the 19 boilermaker helpers were Spanish-surnamed. Of the 3 largest journeyman specialties at the Shops, only the blacksmiths included Spanish-surnamed individuals (2 of the 12), but no Spanish-surnamed apprentices, and only 4 of the 17 blacksmith helpers had Spanish surnames. Thus, among the 177 men known to have been working in the specialties of machinist, boilermaker, and blacksmith, only 42 had Spanish-surnames, and of those, about three-fourths served at the lowest skill and pay level, laborers.

That would change over the years, but for more than a generation Hispanos were woefully underrepresented within the workforce at the Locomotive Repair Shops, which was the only large employer—that is, with more than 500 employees—in the Territory. Overall, in 1900 New Mexico's population was about 60 percent Hispanic. Those Hispanics who did hold jobs at the Shops were systematically marginalized within the skill hierarchy.[9]

The company discriminated even more severely against African Americans. Generally speaking, "Blacks held unskilled jobs such as warehouseman, flagman, messenger, and janitor, as well as menial temporary jobs such as coal shoveler during the rush season."[10] Before the widespread use of mechanized earth-moving equipment, mules were the most common source of tractive power for grading and site preparation, including within rail yards and the grounds of shop facilities. AT&SF, like other railroads, employed many African Americans as mule drivers because they were seen stereotypically as particularly suited to that work.

One of the Spanish-surnamed laborers at the Shops in 1896 was Ignacio Baca, grandfather of former chief justice of the New Mexico Supreme Court Joseph F. Baca and of Mike Baca, a college-educated farmer and vintner from Los Chávez. Ignacio was also progenitor of many other descendants scattered across New Mexico and the United States. Originally from Belen, Ignacio moved in 1880 to 1303 Barelas Road SW, in what would soon become New Albuquerque, specifically to try to get work at the Locomotive Repair Shops. In his middle to late twenties at the time, he was hired on and worked at the shops his whole career, eventually becoming a machinist, retiring with a pension, and helping four of his sons get jobs at the shops. All of them also worked as machinists.[11]

Another early employee at the Albuquerque Locomotive Repair Shops was David Keleher, who was a couple of years older than Ignacio Baca. But Keleher came to Albuquerque in 1881 already a journeyman tinsmith, having completed an apprenticeship in Lawrence, Kansas, where he had been raised by Irish immigrant parents. In Albuquerque Keleher met and married Irish immigrant Mary Ann Gorry. She was working at the time as governess of the children of the Frank W. Smiths. Mr. Smith was general superintendent of the Atlantic and Pacific Railroad. Shortly after their marriage, the newlyweds returned to Lawrence, Kansas, but in 1888 they came back to Albuquerque, and David again signed on with the AT&SF as a tinsmith. The Kelehers lived first at 303 W. Baca (now Santa Fe) Avenue, rented from Santiago Baca, and then from 1893 to 1911 in a house built at the corner of 4th Street and Atlantic Avenue, on a lot purchased from Franz Huning.

The Kelehers' son William wrote many years later,

> The dominant activity of our immediate pioneer neighborhood, and of the little town of Albuquerque, revolved around the Santa Fe Railroad Shops, located at Second Street and Atlantic Avenue, where our father worked until he died in 1903. . . . The neighborhood grew slowly. Many of the settlers were German emigrants who built substantial brick houses and planted gardens. By diverting water in the *acequia madre* running north and south through the area into ditches which they dug adjacent to their homes, they were able to irrigate fruit trees and vegetable gardens. . . . the new town of Albuquerque, depend[ed] upon the railroad shops almost entirely for support and maintenance.[12]

Even well into the twentieth century, almost all of the employees at the Locomotive Repair Shops lived nearby, within walking distance of work, except those who could take advantage of public transportation—trolley or bus—or rode bicycles. That practical choice was reinforced by the AT&SF's policy that mandated that trainmen and enginemen live no farther than three-quarters of a mile from work and strongly recommended that shop employees do likewise.[13] During the

Figure 7.3. Photo of the Albuquerque Locomotive Repair Shops, view from the southeast, ca. 1900. Photographer unknown. Courtesy of the Albuquerque Museum, Photo Archives, Catalog No. PA1978.050.746.

1880s, AT&SF built housing right on the grounds of the Shops and very nearby, which the company rented to employees. Even into the 1890s, some company-owned housing remained at Wallace, Raton, and Albuquerque.[14]

For journeymen machinists, boilermakers, and blacksmiths fortunate enough to obtain employment at the Locomotive Repair Shops during the 1880s and early 1890s, life must have seemed stable, comfortable, and filled with what was considered to be momentously significant work for the country and the world, let alone for oneself and one's family and community of acquaintance. For many, the combination of Albuquerque and work for AT&SF was ideal. People like David Keleher and Ignacio Baca were content to spend their whole working lives at the shops. And they were far from rare among employees at the Shops.

The first dozen years of the existence of the Albuquerque Locomotive Repair Shops coincided with a spectacular period in the business of American railroads, and western railroads in particular. Since the end of the Civil War, with little interruption, there had been a frenzy of constructing railroad lines in the western United States. Scores of railroad companies incorporated, built short runs of track, and sold out for a profit to neighboring lines. Making money by investing in a railroad seemed like a sure-fire way to become wealthy. Wage earners in the Locomotive Repair Shops were, likewise, at the top of the pay scale for manufacturing jobs in New Mexico. Furthermore, as historian James Ducker notes, "As long as a man was white and not a Chicano, he could engage in any type of railroad work and aspire to advance."[15]

By the time the Albuquerque Locomotive Repair Shops were established, the AT&SF had already come to accommodations with the relatively few unions—or brotherhoods as they were known—that were active among railroad workers, though these were limited to the operating crafts—the engineers, firemen, brakeman, and conductors. The repair-shop employees were not directly affected. The shopmen did, nevertheless, benefit from efforts by the AT&SF to provide some benefits to its workforce, put in place to retain trained and effective employees. As Ducker observes, "It was not a great trial for the railroad to recruit men to fill its positions. However, to hire men who would remain faithful and efficient was a challenge the Santa Fe's labor policies worked to meet."[16]

Among such steps taken by the railroad was the establishment in March 1884 of the Atchison Railroad Employees' Association, the goal of which was to "build hospitals and contract with doctors throughout the line to care for sick and injured employees."[17] This mirrored the 1882 building of a railway workers' hospital at two-year-old New Albuquerque by the AT&SF subsidiary the Atlantic & Pacific Railroad. Although originally provided as a free service, beginning in 1884, the company deducted from workers' wages to support the hospital system. By 1894, the monthly payroll deduction throughout the AT&SF system for support of hospital care ranged from thirty cents to one dollar, depending on pay rate.[18] The A&P also paid for burial space at Albuquerque for employees who died on the job and had no known family.[19]

In addition, during the 1880s the Santa Fe set up "reading rooms" in Albuquerque and elsewhere across the Southwest. Each reading room made available copies of magazines and books, as well as space for playing cards and billiards. The amenities also generally included toilet and bathing facilities because many of the male AT&SF employees were single, ate at restaurants, and slept in a variety of temporary accommodations. Thus, keeping clean was often difficult and therefore neglected.

Looking back from 1902, AT&SF president Edward Ripley wrote about the reading rooms that "the Santa Fe management had in view several things: (a) To aid the employes [*sic*] and their families in self-development. (b) To surround them with influences by which their lives would be brighter and more hopeful. (c) To give them an opportunity of making themselves worthy of promotion to higher spheres. (d) To put a new value on a man's life, and emphasize brain and conscience power as a factor in railroad operation."[20] Responding to the trend of the times, Santa Fe employees of the late nineteenth century "showed an abiding drive to improve themselves."[21] Many, especially shopmen, were prolific inventors and refiners of tools, equipment, and methods.

Certainly, relations between shopmen and their supervisors were not always congenial. Shopmen at Albuquerque, as at other locomotive repair shops, frequently complained about arbitrary and overly stringent penalties imposed by supervisors for honest mistakes or because of animus toward a particular employee. After years of complaints to management about abuse of shopmen by their immediate supervisors,

> On August 1, 1897, the Santa Fe adopted a discipline program . . . [based on] a scheme of merits and demerits called the Brown system, after its originator, George R. Brown, general superintendent of the Fall Brook Railroad in New York. . . . Under the system the company could still hand out summary discharges for major offenses, but all others could be punished only by reprimands or demerits, a record of which was kept for each employee in the offices of the division superintendent. An employee was notified of each mark entered against his name and had the right to appeal those he considered unjustified. Should too many demerits, or "brownies" as they were commonly called, appear beside the worker's name the superintendent could call on him to explain the mistakes, and, if unsatisfied, dismiss him.[22]

The contracts that the company offered to shopmen in the 1880s were considered to be as generous as any in the industry. In the middle of the next decade, skilled shopmen earned $2.50 to $3.00 a day working for AT&SF, with shop foremen making more than $3.00.[23] Nevertheless, as we will see in chapter 10, AT&SF shopmen walked off the job in the early 1890s and forced the company to sign contracts with the shop craft brotherhoods that included pay increases. That seemed only fair with railroad earnings leading the national economy to dizzying heights.

As in economic bubbles familiar from more recent times, the price of railroad stocks climbed and climbed with no end in sight. Until that end came, rapidly, late in 1892 and the first months of 1893. At that time, the collapse and bankruptcy of multiple railroads led the way into what has been called the "depression of the nineties," which also strongly affected manufacturing, agriculture, and construction. The resultant unemployment was at least on a par with that experienced during the Great Depression of the 1930s.

Overly optimistic and exuberant expansion of AT&SF's network of track was both symptomatic of and contributory to industrywide economic profligacy. In 1890, the company spent extravagant sums of money to buy two railroads of highly questionable worth: the Colorado Midland and the St. Louis and San Francisco.[24] Weighed down with debts from these purchases and other similar unwise expenditures, the Santa Fe was auctioned off to a receiver. That cast a pall of uncertainty over the workforce at the Albuquerque Locomotive Repair Shops and throughout the Santa Fe system as a whole.

At almost the same time, two other important railroads also failed, the Northern Pacific Railway and the Union Pacific Railroad, plunging the entire industry nationally into a five-year-long depression. The AT&SF went into receivership in December 1893 and did not extricate itself for two years.[25] Even then, growth of all railroads remained sluggish for years. One economic historian who specialized in this period sums up the sluggishness this way: "The depression which started in 1893 was not fully overcome until after 1900." Again writing about the national situation of railroads, he wrote, "From 1892 to 1895 gross new railroad mileage dropped from 4,584 miles to 1,938 miles, a decline of about 58 per cent. Not until the turn of the century did new railroad mileage again attain the 1892 levels." Likewise, purchase of steam locomotives by all US railroads combined fell from a peak of 205 per quarter in 1892 to a low of less than 20 per quarter.[26]

Nationwide, unemployment peaked at about 20 percent during the 1890s. With flattening of the trend in growth of freight and passenger miles recorded by the AT&SF during this period, the workforce at the Albuquerque Shops grew at a much slower pace than it had during the previous decade. But we do not find evidence of layoffs or furloughs of any significant duration at the Albuquerque Shops during the depression of the nineties. Nevertheless, the mood among railroad workers was one of anticipating the hitherto unthinkable demise of American railroading.

CHAPTER 8

Work Schedules and Routines

During the late nineteenth and early twentieth centuries, the rhythm and pace of life at the Locomotive Repair Shops, in the neighborhoods of Barelas and South Broadway and throughout much of the rest of Albuquerque, were governed by the routines of steam locomotive overhaul. Those routines began before most of the workers arrived in the morning. The stationary engineers fired up the boilers that powered electric generators to supply most of the smaller power tools and to provide light in the Shops. Separate boilers generated steam that was conducted by pipes and heavy hoses to numerous locations for use in cleaning parts and hosing down floors and walls. Other water heaters supplied the four large washrooms that allowed workers to scrub off the grease and grime that came with locomotive work. During the cold season, massive heating coils on the roofs of major buildings and the associated distribution fans had to be switched on so that by the start of work interior temperatures would be sufficiently warm. When the weather itself was warm, the same fans were turned on to force cooler air through ductwork to help cool the massive Shop spaces. An August 1922 article enthused that "[this] ventilating system has a capacity sufficient to change the air in the entire shop building three times in an hour."[1] Furthermore, 4 percent of the glass panes making up the walls of the machine and boiler shops were hinged, allowing them to open with a series of mechanical cranks to permit cooler exterior air to flow into the buildings or hotter air to rise out.

Figure 8.1. Shopworkers washing up at the end of the day's work at one of four lavatories at the Albuquerque Repair Shops, 1943. Photo by Jack Delano. National Hist Reg Nomination, 2014, figure 12.

Under normal circumstances, a steam whistle sounded at the Shops every working day at 7:45 a.m. As Eloy Gutiérrez, an apprentice sheet metal worker in the early 1940s, remembered, he would get out of bed as soon as he heard that whistle, throw on his coveralls, and reach the time clock before the whistle sounded a second time at 8:00.[2] That time clock, in a small building near the entrance gate, was one of several scattered around the shop grounds. Because the lunch hour was not paid time, employees were expected to clock out at noon and back in at 1:00. Work in the Shops involved routinely handling greasy, grimy parts and equipment; therefore, shopmen habitually scrubbed up in one of the industrial lavatories at lunchtime and before leaving work in the afternoon.

Figure 8.2. Shopworkers checking out for lunch at one of the time clocks in the Albuquerque Locomotive Repair Shops, 1943. Photo by Jack Delano. Courtesy of the Library of Congress, catalog number LC fsa 8d15501.

There were commercial laundries scattered on streets around the Shops to deal with the mountains of oil-caked work clothes of shopmen. For instance, there were businesses such as the Albuquerque Steam Laundry, which later became the long-lived Imperial Laundry at 2nd Street and Lead Avenue, and the Hubbs Laundry, which became the Excelsior Laundry at 2nd Street and Coal Avenue (later moving to 2nd and Roma).[3] In addition, there were smaller laundries, which in the nineteenth century were often run by Chinese entrepreneurs, such as Chong Lee at 307 S. 2nd Street, Hong Gee at 409 W. Central, and Hop Sing at 214 N. 3rd Street.[4]

Even with all the scrubbing of hands and arms and faces, and the boiling of clothes, it was sometimes not possible to rid oneself of all the

Figure 8.3. Photo of the Albuquerque Steam Laundry on S. 2nd Street, with employees, 1880s. Photographer unknown. Courtesy of the Albuquerque Museum, Photo Archives, Catalog No. PA1978.050.743.

noxious residue of work in the Shops. Former chief justice of the New Mexico Supreme Court Joseph F. Baca, who had many relatives who worked at the Albuquerque Locomotive Repair Shops, remembered one of his uncles in particular, a machinist, who no matter how clean he looked always seemed to smell of oil and grease. It was as though the various lubricants suffused his entire being.[5] There were other shop-men, though, who in their off-hours managed to betray little evidence of their constant daily contact with coal dust, petroleum products, paint, and the fine residue of shaping and polishing metal. Bonifacio Shaw, for instance, a machinist at the Shops in the 1920s and 1930s, although he wore overalls to work, at home he "always had clean hands, and on the weekends he wore a nice, freshly laundered shirt, and a tie."[6] Some shop employees had working conditions that were generally less grimy than those of their workmates, Joseph Swillum for one. During his more than forty-year career as an apprentice instructor, he almost always wore a tie, coat, and hat to work. Of course just being

Figure 8.4. Photo of safety warning sign, depicting the character "Acci Dent," stenciled on a door at the Albuquerque Locomotive Repair Shops, undated. Courtesy of the photographer, Patrick Trujillo.

on the grounds of the Shops and meeting other employees meant that sometimes his wardrobe ended up less than spotless.[7]

Many shop employees walked or rode bicycles to their nearby homes for lunch. Others carried their lunches to work in lunch boxes or bags, which during the morning were stashed in lockers that were assigned to each individual. Some even went to cafes and lunch counters, such as several on 1st Street where a person could eat for twenty-five cents.[8]

The lunch hour on Thursdays was different, as attendance at a safety meeting was required. The company used images of a cartoon-like figure called "Acci Dent," painted on walls and other surfaces around the Shops, to remind employees of dangers and precautions, and they were admonished about these at the Thursday lunchtime meetings. Avoiding accidents and injuries was very important to most shopmen because, as Eloy Gutiérrez said, if you got "injured, you got fired."[9] Nor was there any sick leave, so most men worked even when they were not well unless the illness laid them up.[10]

Another shrill of the steam whistle marked the end of the workday. Ordinarily, that was at 5:00 p.m. The powerful whistle could be heard all over Albuquerque, so it synchronized many people's activities, whether they were directly connected to the Shops or not. Judge Joseph Baca remembered that, as a youngster, when he heard the 5:00 o'clock whistle, he would run inside the house in order to listen to *Superman* on the radio.[11] The whistle also sounded at times of emergency and on special occasions. In mid-May 1941 and again a year later, for example, extremely heavy rain swelled the Rio Grande. When the river's waters threatened to flow into city streets and drown the zoo, the steam whistle at the Shops sounded the alarm.[12] Calling up a very different emotion, attorney Michael Keleher remembered vividly that he was in Presbyterian Hospital as a child on the morning of April 6, 1945, having his tonsils removed, when at 6:00 a.m. the whistle at the Shops began to blare. It continued for many minutes, announcing the end of World War II in Europe, or V-E Day.[13] When he was ten to twelve years old, Joseph Baca was stationed outside one of the shop gates at quitting time, as a paper boy handing out the newsletter *El Greco*.[14]

Prior to the 1916 and 1917 passage of federal laws and regulations governing the standard workday for railroad employees, the ordinary workday was nine hours long, and sometimes longer still.[15] The work week, too, was longer, typically including Saturday. During both World War I and World War II, as well as during the heart of the Great Depression, the work schedule at the Shops varied considerably from that of more normal times. As former machinist apprentice Mike Baca said, during World War II, the workday at the Shops began at 7:00 a.m. and was nine hours long.[16] The Shops also operated seven days a week during the war. That did not mean that individual shopmen worked nine hours a day, seven days a week, although that was not unheard of. And Olivia Cordova Loomis recalled that in 1962 or 1963 the company changed the daily work schedule by decreasing the time for lunch to just half an hour. Thus, the workday ran from 7:30 a.m. till 4:00 p.m., including the lunch period.[17]

At the opposite extreme of what was required during the World Wars, during the Depression, Thomas Cordova and his fellow shopmen

Figure 8.5. Shopworkers leaving the grounds of the Locomotive Repair Shops by the west gate at the end of the day, 1943. Photo by Jack Delano. Courtesy of the Library of Congress, catalog number LC fsa 8d15944.

were repairing locomotives at the Shops only "maybe three days a week." Nevertheless, Cordova was proud years later that he had still been "able to pull his family through."[18]

After hours, many shopmen pursued activities that kept them in the company of other Santa Fe workers. The reasons behind that were not at all difficult to imagine because the residential areas encircling the Shops seemed to be home mostly to AT&SF employees and their families. For example, industrially sponsored adult amateur baseball has long been popular in the United States: "By the 1890s . . . shopmen throughout the [AT&SF] line annually organized baseball teams to play each other and whatever other competition could be found."[19] Many Albuquerque shopmen joined their coworkers to play baseball in

Figure 8.6. Photo of the Albuquerque Shops' All-Star baseball team, near Tingley Field (formerly Apprentice Field). The Sandia Mountains can be seen dimly in the distance. Undated. Photographer unknown. Courtesy of Center for Southwest Research, University Libraries, University of New Mexico; H. W. Stowell Pictorial Collection, Box 1, PICT 000-505-0095.

the evenings during spring, summer, and fall. Bonifacio Shaw, for example, was always on the baseball diamond after work.[20] Albuquerque machinist Amado Baca played ball for the Barelas Blues in the 1930s.[21] When they were not playing baseball, many shopmen were in the stands watching baseball games involving local teams. And when the Brooklyn Dodgers moved to Los Angeles in 1958, a number of Albuquerque shopmen began using their railroad passes to ride the Santa Fe to Southern California from time to time to watch a game or two.

Recreation, though, was not the only way shopmen used their off-hours. Eloy Gutiérrez knew one young shopman who spent his evenings as a carpenter, building apartments, and "worked himself to death."[22] Quite a number of shopmen cultivated vegetable gardens in their spare time, thereby simultaneously continuing family agrarian traditions and adding to the wellbeing of their own families.

The company, reflecting generally paternalistic attitudes of employers of the day, worried about young single employees getting into trouble when they were away from the structure provided by their jobs. In Albuquerque the AT&SF donated land for a YMCA (the Y) building on Central Avenue. The local Y became a joint city and railroad enterprise. With similar reasons, the Santa Fe provided space on the third floor of the General Administration Building at the Shops as a reading and billiards room. As explained by the Santa Fe's superintendent of reading rooms, "The general plan contemplates having one large room for a reading room in which are placed all the periodicals of the day, and a library of choice books in circulation, a smaller room for cards and amusements, a large room for billiards, also bath rooms [sic], toilet and wash rooms. Recently, we have added sleeping rooms and bowling alleys. The use of these privileges is absolutely free to all employees regularly on the payroll of the company."[23] "As elsewhere within the AT&SF system, the Albuquerque 'Y' provid[ed] services similar to the reading rooms, . . . held socials, Bible studies, and prayer and inspirational meetings."[24]

In addition, there was a club especially for apprentices and their families, which met on Saturday evenings in the 1920s and later at the El Fidel Hotel in the 500 block of Copper Avenue NW. Usually also in attendance at the club meetings were the apprentice instructors.[25] Furthermore, all members of the shop crafts brotherhoods had union meetings as often as twice a month, as well as annual national meetings that some local union officials would attend.

Once shopmen reached journeyman status at the Albuquerque Repair Shops, they tended to stay put for a long time. Many worked their whole careers at the Shops, thirty, forty, even fifty years. Thomas Cordova, for example, retired in 1976 after forty-nine years, including a couple of layoffs. His railroad pension was good enough that it paid

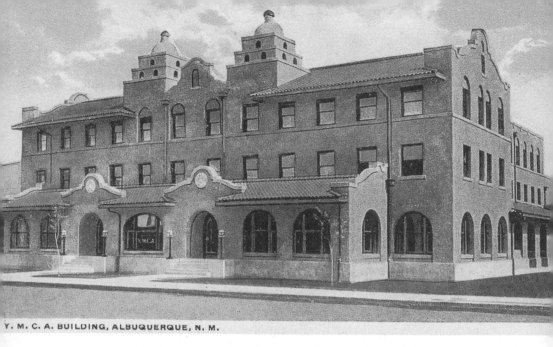

Y. M. C. A. BUILDING, ALBUQUERQUE, N. M.

Figure 8.7. Postcard depicting the Albuquerque YMCA building, at the corner of Central (formerly Railroad) Avenue and 1st Street, early 1900s. Authors' collection.

Figure 8.8. Photo of retirement party at the Albuquerque Repair Shops for Frank G. Gómez, sheet metal worker, and Raven Bean, machinist, after thirty-six and thirty-five years of service, respectively, 1958. Gómez with placard; Bean to his right. Courtesy of Center for Southwest Research, University Libraries, University of New Mexico; H. W. Stowell Pictorial Collection, Box 1, PICT 000-505-0112.

him more money in retirement than he had earned in some years he worked.[26] Carman and welder Louis Johnson worked thirty years for the AT&SF, part of that time at the shops in Richmond, California, retiring from the Albuquerque Shops about 1975.[27] Having started working as an instructor of apprentices at the Albuquerque Repair Shops in 1914, Joseph Swillum retired in 1955, just as the transition to diesel power was being completed. He had put in a full forty years with the railroad, but he kept golfing another twenty-two years.[28] Refugio José Baca hired on as a machinist apprentice at the Shops in 1907 and retired as a journeyman machinist forty-five years later, again as steam locomotives were being retired en masse.[29] Thomas G. Turrietta worked as a carman at the Albuquerque Repair Shops for thirty-nine years.[30] One thing suggested by all this information on length of service is that many Albuquerque shopmen must have been generally content with the working environment there and with their retirement pensions. It was very common for the retirement of long-time Locomotive Repair shopmen to be marked with a brief party at the Shops.

CHAPTER 9

Albuquerque and the Locomotive Repair Shops

1901–1922

The years between 1901 and 1922 were momentous in a number of ways for workers at the Albuquerque Locomotive Repair Shops and for the Atchison, Topeka & Santa Fe Railway as a business. As historian Richard Frost notes, it was "the dramatic rise of Los Angeles [California] in the late 1890s and the early 1900s [that] made the Santa Fe a major transcontinental railroad."[1] AT&SF annual reports show that from 1900 through 1915 total annual operating revenue (freight and passenger) for the company jumped by almost two and a half times, from just over $48 million to $117.6 million. In 1915, the company proudly announced to its shareholders, "The year has been the largest as to earnings, both gross and net, in the history of the Company." The management, considering only the most recent year, attributed the rise in earnings to "the unprecedented wheat crop of Kansas and the largely increased yield of agricultural products of all kinds in the so-called 'Plains Country.'"[2] The trend of earnings, though, had been upward in almost every one of the preceding sixteen years, suggesting that other factors besides just a single year's bumper crop were multiplying business for the railroad. The mushrooming demands of Californians for manufactured goods as well as spectacular production of fresh fruits and vegetables that could now be shipped eastward in record time in refrigerated cars undoubtedly played a major part, as Frost suggests. Despite record income, the company complained about the shipping revenue lost because of the Panama Canal, which had just opened in August 1914. This again suggests the importance of California to the railroad, both as a shipper and as a destination for freight and passengers.

Despite record earnings, competition from ocean transport by way of the Panama Canal was not the only circumstance that AT&SF management publicly lamented. The cost of repairing locomotives and of maintaining and improving shops was rising alarmingly. Corporate annual reports for 1905 and 1915 recorded an increase in the operating costs of repair shops systemwide from $5.38 million to $8.1 million, a jump of nearly 51 percent.[3] Naturally, wages constituted the major shop expense, which prompted this statement of grievance on behalf of the company in 1914: "As a class they [our employees] are a credit to themselves and the road. Left to themselves there would be little of which to complain, but the [labor] organizations as a body have been aggressively demanding increased wages for their members with no regard for the ability of their employers to pay, and have been steadily demanding, and frequently with success, many varieties of legislation."[4]

Dramatically increased railroad traffic directly impacted work at the Albuquerque Locomotive Repair Shops.[5] The number of AT&SF steam locomotives operating at the end of June 1915 was 2,105, up from 1,454 just ten years earlier, almost a 50 percent increase.[6] That meant correspondingly more overhauls and more routine repairs had to be performed. As we will see in chapter 11, the dramatically increased workload, coupled with larger and larger locomotives, necessitated enlarging the Albuquerque Shops and significantly expanding the number of shopmen in all categories.

This report from October 1903 is one indication of the increasing size of AT&SF locomotives in the early 1900s: "The largest engine in the world, #989, which has been in the [Albuquerque] back shops [a term in general use for repair shops] being almost entirely reconstructed, will be turned out of the shops next week."[7] In 1903, Santa Fe #989 was an almost new locomotive with a 2–10–0 wheel configuration (with ten drive wheels), known as a decapod, a type particularly suited to mountainous regions. Another locomotive of the same class, built almost sixteen years later, was 71 feet long, including its tender, and weighed more than one hundred tons, which gives an approximation of the size of #989.[8] The shopmen's pride in having almost completely rebuilt the world's then-largest locomotive is evident even in this matter-of-fact news note. A locomotive of that size would have just barely fit in the erecting bay of the old stone-walled machine shop, which was only 74 feet 10 inches wide. Even with its tender off, that would have left precious little maneuvering room around the locomotive.

At the same time that the rebuilt #989 was to be released for a test run, "the roof over the machine shops [was] nearing completion," representing reconstruction after a recent fire. Such catastrophes were not unexpected at railroad repair shops, given the constant proximity of volatile substances, very hot metal, welds, frequent sparks, and reliance on wood-timber roof structures. In fact, in July 1904 and again in July 1922, the machine shop and much of the rest of the AT&SF–Gulf Lines locomotive repair complex at Cleburne, Texas, were consumed by huge fires.[9] The Cleburne shops had exhibited many of the same characteristics that made the pre-1920s Albuquerque Repair Shops vulnerable to fire, including timber-frame roof structures. When a new shop complex was built at Cleburne between 1926 and 1930, it followed plans, methods, and materials very similar to those pioneered at Albuquerque in putting up the "state of the art" Shops from 1914 to 1924 (see chapter 11).

At Albuquerque itself, AT&SF made many adjustments and improvements to the Shops complex in addition to the wholesale rebuilding of the early 1920s. The diameter of the turntable, for instance, was lengthened twice during the early twentieth century, increasing from 54 feet in 1900 to 85 feet, and then to 120 feet in 1914. Because of the increasing length of locomotives, even that was not the final turntable diameter. In 1942, it was increased an additional 20 feet. Larger and larger boilers in locomotives also meant a need for larger and larger water storage tanks within the Shops complex—and more water wells to keep up with the burgeoning consumption of water at the Shops.[10]

All these developments brought with them a need for an increased workforce at the Shops. According to the 1910 Census, the number of hourly shopmen at Albuquerque was then under 500, but war in Europe stimulated train traffic in the United States, which relentlessly pushed the number of shopmen higher. In 1920, the number of hourly shopmen was 970. Including salaried shop staff, the city directory for 1919 indicated total employment at the Shops for that year of 1,195.[11]

One of the employees who joined the workforce of the Albuquerque Locomotive Repair Shops during the first two decades of the twentieth century was Joseph Swillum. He came to Albuquerque, by train naturally, in 1914 at age twenty-three with the University of Missouri glee club on a tour to California. One of the members of the singing group came down with smallpox upon arrival in Albuquerque, and as a result, the whole glee club was quarantined and could not continue the tour.

During his enforced stay in Albuquerque, Swillum became entranced by the growing city and its surroundings. He got to know people at the Shops and, as a recent college graduate, was offered and accepted a position as a special apprentice on July 6, 1914. His beginning pay was $1.75 a day. From then until 1957 he worked as an instructor at the Shops, guiding more than 1,500 AT&SF apprentices through the basics of math and mechanical engineering. From 8:00 a.m. till noon every workday, Joe prepared his lessons. After a four-hour break, he began his teaching day in the general office building, giving classes from 4:00 to 9:00 p.m.

The midday break suited Joe just fine. He became an avid and life-long golfer, spending his Wednesday breaks during good weather at the Albuquerque Country Club, of which he was a founding member. He married Magdeleine Heibel in 1918, and they lived in an apartment on 3rd Street until 1927, when they purchased a house at 416 13th NW. The couple became good friends of the family of fellow AT&SF employee Amado Baca, as well as the grandchildren of former shopman David Keleher.[12]

Bonifacio Shaw, whose mother was from Isleta Pueblo, began work as a machinist at the Shops in about 1906, eight years before Joe Swillum. Bonifacio had a long career with AT&SF, retiring about 1960.

He was both a player and a fan of baseball. As a perk of employment with the Santa Fe, as it was for all AT&SF employees, he had a pass for free travel on the train. He would frequently ride the train to Los Angeles, leaving Albuquerque late on a Friday. He would attend two Dodgers' games over the weekend and be back in Albuquerque by Monday morning.

At noontime on workdays, Bonifacio would walk home from the Shops to 923 S. Arno for beans, red chile, and tortillas with butter, his usual lunch. His grandson Patrick Trujillo remembers that when Patrick was a young boy his grandmother, Margaret Armijo Shaw, would send him outside to sit on the fence to watch for when Bonifacio reached Broadway on his way home at midday. That way Patrick could run and tell his grandmother so that she would have Bonifacio's lunch hot when he arrived.[13]

As time went on, the payroll at the Shops kept growing, and expansion at the Shops drove a similar increase in Albuquerque's population. The 1920 census reported that Albuquerque was then home to 15,157 people, up 66.3 percent since 1910 and up an astounding 143 percent since 1900.[14] By 1919, Albuquerque exhibited many of the characteristics of twentieth-century American urban life. There were, for example, thirty churches in Albuquerque: seven each that were Catholic and Methodist Episcopal; two each were Baptist, Christian Science, Episcopal, Lutheran, and Presbyterian; one each were Christian, Congregational, Church of Christ, Seventh Day Adventist, and a Gospel Hall—and there was a Jewish synagogue. The YMCA, at the corner of Central Avenue and 1st Street, was a frequent after-hours haunt for shopmen. On account of the Y's presumed beneficial effect on employees' spirits and behavior, it enjoyed financial support from the Santa Fe.[15]

With AT&SF's presence and permanence in Albuquerque established, other large industrial enterprises were able to capitalize on the continental reach of the railroad:

Soon after the turn of the [twentieth] century, a second major industrial employer arrived in the city—the American Lumber Company—whose fortunes were also tied to the railroad.

Incorporated in 1901, the company purchased timber lands in the
Zuni Mountains, some 100 miles west of Albuquerque, and in 1903
the company was ready to build a sawmill and associated wood-
working factories on 110 acres of former agricultural land just
northwest of the city limits. The sawmill plant and other buildings
were connected by a railroad spur to the AT&SF's main line. This
allowed easy access for incoming shipments of logs cut in the white
pine forests, and then a convenient shipping method to markets
throughout the West. By 1908, the American Lumber Company
was reportedly the largest lumbering enterprise in the Southwest. It
was comprised of sawmills, a box factory, and a sash and door factory, large holding ponds for unprocessed logs, and its own electric
plant. . . . Thirty to forty carloads of logs were shipped from the
plant every day and as much as fifty million board-feet of finished
lumber were produced per year. Within three years of opening its
mill, the company employed more than 850 people.[16]

For a long while the American Lumber Company was Albuquerque's
most financially significant enterprise after the Locomotive Repair
Shops. American Lumber was the city's second largest employer until
1914, when it first ran into financial difficulty. After going through
several reorganizations and name changes, the much reduced lumber
company, finally closed its doors in 1942.[17] There were many other,
smaller manufacturing businesses and wholesale distribution companies that also relied on AT&SF track and rolling stock beginning as
early as the 1880s and proliferating in the twentieth century.

During the first two decades of the 1900s, the number of visitors to
Albuquerque surged, in significant part because of promotional activities undertaken by AT&SF and the Fred Harvey Company, which
served as the hotel and restaurant concessionaire for the railroad. In
1902, the Santa Fe built the mission-style Alvarado Hotel, the system's
largest hotel, as well as an Indian Curio Building and a new passenger
depot at the north end of the Albuquerque rail yard complex.[18] It was
the exoticism of Pueblo and Hispanic customs and arts, the ruins of
ancient ancestral Puebloan "apartment-towns," and New Mexico's

Figure 9.1. Postcard depicting the American Lumber Company complex in northwest Albuquerque, ca. 1908. Authors' collection.

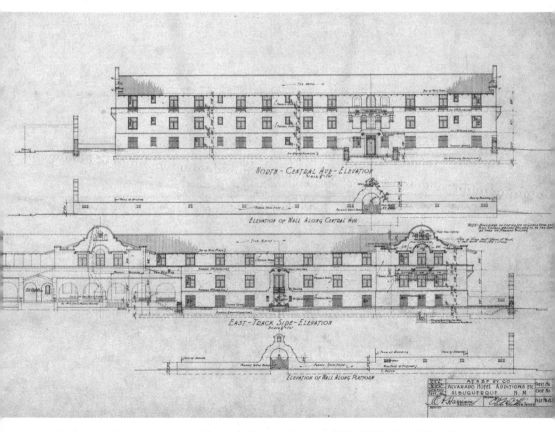

Figure 9.2. Architectural drawing of the north and east elevations of the AT&SF's Alvarado Hotel in Albuquerque located just north of the AT&SF depot, made in preparation for enlargement of the hotel, 1932. E. A. Harrison, architect for AT&SF. Courtesy of the Kansas State Historical Society. DaRT ID: 57312.

chiseled and colorful landforms that drew increasing numbers of tourists, who came mostly by rail. In contrast, only a half mile south along the tracks from the AT&SF depot and the Harvey Indian Curio Building stood the Locomotive Repair Shops, the sprawling, towering, busy epitome of the "modern" Machine Age.

The early years of the twentieth century saw Albuquerque emerge as a destination for the treatment of tuberculosis because of its high, dry, and temperate climate. Because of that "several sanatoriums were erected in the sand hills east of the railroad." Among them were St. Joseph's Sanatorium, established by the Sisters of Charity in 1902, and the Southwestern Presbyterian Sanatorium, which opened its doors in 1908. Patients at Presbyterian included [future senator] Clinton P. Anderson, who by 1919 was working as a journalist for the *Albuquerque Morning Journal*.[19]

The population continued to grow as people from other areas of the country moved West. As historian Erna Ferguson remembers, "By the turn of the [twentieth] century, . . . The railroad payroll kept the wheels turning. Trade by rail was just as profitable as trade by wagon, and men found work along the tracks and in the shops. Irishmen and Italians had swelled and enlivened the body politic."[20]

To accommodate single shopmen, by 1919 at least thirteen apartment buildings and boarding houses were in operation in Albuquerque in the vicinity of the Shops. For much the same reason, there were at least fourteen restaurants and lunchrooms offering daily meals within easy walking distance of the shops. Many employees with families owned their residences. For example, the 1919 city directory lists the top management at the Shops as David E. Barton, superintendent of shops; John P. McMurray, master mechanic; Charles F. Stucke, general shop foreman; Aubrey B. Wachter, storekeeper; Martin T. Murphy, chief clerk at the supply depot; and Ralph P. Brown, electrical foreman. Of these six managers, four lived in their own residences, suggesting that they probably had families. The remaining two, Wachter and Brown, rented rooms indicating that, like many other shopmen—especially apprentices and helpers—they were single.[21]

The following excerpt from the "Railroads and Shops" column in

Figure 9.3. Photo of the Red Ball Café on 4th Street SW in Albuquerque, date unknown, 1922 or later. Located three blocks southwest of the west gate of the Shops, it has long sold Wimpy burgers. Photographer, unknown. Courtesy of the National Hispanic Cultural Center, Library/Archive, photo #501, Box 2, Folder 3.

the October 15, 1903, issue of the *Albuquerque Morning Journal* provides a glimpse of typical happenings in the Shops during the early twentieth century: "Both dynamos [electric generators] were running yesterday for the first time since the fire [evidently referring to the fire mentioned earlier that destroyed the machine shop's roof]. Heretofore the machinery throughout the shops has only been operated by one dynamo, but since the other one has been repaired the motive power has doubled.

The car repairing yards have been showing a remarkably clean appearance for the past few days. The force have been working hard for the past month in order to clean up with the back work, in order to have a few days vacation during the week [of the territorial fair].

All the shop employes [*sic*] will have a day off today. The entire shops will be shut down, except the portions of the roundhouse, which are on constant duty.

A. Rumberg, who has recently been promoted to boss of the machine shops, met with quite a painful accident yesterday, from which he will be laid up for several days. Mr. Rumberg, while carrying a driving box [an iron or steel casting that holds a brass or bearing for a drive axle], accidentally let it drop on his feet, completely mashing a toe on his right foot. He will not be able to return to work for a week or more.

Engine 822, which was damaged in a recent wreck, was turned out of the back shops yesterday, after a thorough overhauling. The engine will take a trial run to Isleta and back today.[22]

Those were some of the ordinary, daily, local challenges, but wide-scale difficulties could also affect all customers and employees of the Santa Fe. The year 1905, for instance, "was remarkable for excessive rainfall not confined to any one locality, but almost universal and nearly continuous. . . . On at least three occasions your [referring to the stockholders] main lines in Arizona and New Mexico were totally disabled for from four to eight days, besides innumerable smaller breaks. For weeks it was necessary to advise intending patrons to ship or travel by other routes. . . . The cost of repairing the damages caused by the floods will amount to $2,000,000."[23] Such interruptions of service necessarily also blocked the delivery of parts, lubricants, and raw materials to the Albuquerque Locomotive Repair Shops, putting a crimp in the overhaul schedule.

CHAPTER 10

The Railroad Shopmen's Strikes of 1893 and 1922

I n the late nineteenth century, to many in the rest of the United States, the Rio Grande Valley of New Mexico, a recently annexed Mexican territory, seemed alien and extraneous to developments in the nation. Likewise, in the 1880s and 1890s, from the perspective of most New Mexicans, the bulk of US land and people had very little relevance to everyday life in the former Mexican province.

The explosion of railroad building that laced New Mexico during the last two decades of the nineteenth century, however, tied the territory in unanticipated ways to social, political, and economic events and frameworks that originated east of the Mississippi. Occurrences far distant from the US–Mexico borderlands could suddenly turn New Mexican lives upside down. In particular, the rapid industrialization of the eastern United States following the Civil War brought on a parallel rise of the industrial labor movement.

Large-scale industrial work established conditions under which employers were able to exercise unprecedented control over workers' lives, "where [they] resided, how frequently [they] moved, how much time [they] had for [their] famil[ies], how comfortable a life-style [they] could afford, and how long [they] might live."[1] Many workers formed organizations aimed at blunting the extraordinary power employers held over their lives. Beginning at the level of single local businesses, employee organizations sought some say-so over their members' hours, pay levels, and working conditions. Fierce resistance from some employers to what they saw as their employees wantonly ganging up on them sparked efforts to strengthen worker organizations by banding together with other similar groups. The resulting fraternal craft associations were not altogether new, harkening back all the way to medieval European guilds.

Tit-for-tat countermeasures by employers and workers resulted in allied crafts—for instance, railroad engineers and firemen—combining their associations into ever larger entities. In 1886, the American Federation of Labor (AFL) came into being as a national alliance of fraternal craft organizations. The railroad shop crafts were slower to organize for collective bargaining.

A heavy anti-union sentiment long dominated both the New Mexico territorial-state government as well as the management and ownership of the Santa Fe, which repeatedly and regularly railed against the "excessive" demands of railroad workers' brotherhoods. The company over and over blamed unions for threatening its economic survival. This statement from the company's annual report for 1914 is typical: "[AT&SF] employes [sic] . . . are a credit to themselves and to the road. Left to themselves there would be little of which to complain, but the [labor] organizations as a body have been demanding increased wages for their members with no regard for the ability of their employers to pay."[2] Nevertheless, "in 1892 the Santa Fe [had become] the first [railroad] to sign contracts with the International Association of Machinists, the International Brotherhood of Blacksmiths, the National Brotherhood of Boilermakers, and the Brotherhood of Railway Carmen."[3] Included in those contracts were provisions for wage increases and improved working conditions for shopmen represented by those organizations.

Only a year later, when convulsion in the railroad industry opened what has come to be called the depression of the nineties, AT&SF reneged on the shopmen's contracts, and the workforce at the Albuquerque Locomotive Repair Shops—as well as New Mexico in general—experienced outsized seismic shocks. In mid-April 1893, shopmen throughout the Santa Fe system walked off the job. In New Mexico that included especially shopmen at Raton, Las Vegas, and San

Marcial. By April 18, Santa Fe trains had been abandoned at several places in Kansas because of a lack of locomotives in safe running order. And non-union replacement workers at La Junta, Colorado, had joined the strike of union shopmen there. Even though work stoppages were not reported at Albuquerque, train traffic throughout the Western Division of the AT&SF fell off drastically. The virtual absence of strike activity at Albuquerque may have been due to the fact that Albuquerque was headquarters for the division and therefore was home to many members of the company's management, who were able to prevail on the shopmen there to continue working. On April 25, the *Las Vegas Daily Optic* reported, "The strike on the Atchison is ended."[4] The decisive factor in ending the strike in favor of the shopmen was the inability of the AT&SF to recruit and hold a sufficient number of replacement workers to keep locomotives serviced, while perishable cargos began rotting in box cars.[5]

For the general public, both in New Mexico and in the nation as a whole, the 1890s depression began with a horrifying rise in bank and business failures, including in February 1893 the bankruptcy of the Philadelphia and Reading Railroad. That would be only the first of many railroad failures. By the middle of 1894, 156 railroads had failed, including the Atchison, Topeka & Santa Fe Railroad. Further, the national unemployment rate was running at nearly 20 percent.[6] Business in all fields declined, including in the transportation of goods and people by rail. As we saw in chapter 7, the AT&SF went into receivership in December 1893, even though the shopmen's strike had ended and full operation had been restored before the end of April of that year. Of all the railroads that failed that year, "the Santa Fe was the largest in track miles—4,483—and the largest in capital investment: $228 million in bonds and $102 million in stock, or a total amounting to a third of a billion dollars."[7]

In December 1895, a purchasing committee bought the Atchison, Topeka & Santa Fe Railroad Company at public auction, comprising all its track, rolling stock, facilities, and other property, including the Albuquerque Locomotive Repair Shops. With a slight name change—from Railroad to Railway—the purchasers reorganized the company without causing interruption in rail service or work by employees. During its first six months of existence, the new company spent about $1.1 million on shops and shopwork systemwide. As of June 30, 1895, the AT&SF owned 839 locomotives subject to periodic maintenance in its shops.[8]

Over the next twenty-five years, the number of AT&SF locomotives rose to 2,195, an increase of about 160 percent.[9] It was two and a half decades of spectacular, if uneven, growth for the company. It was also a period during which the managements of many railroads took steps to prevent a recurrence of their feeling of impotence during the 1893 shopmen's strike. The next time, they thought, they would be ready with overwhelming force to put down any similar labor action by the railroad crafts. And, indeed, in the very next year of 1894 the Pullman Strike was violently put down.

But the railroads could not anticipate the national political shift of US entrance into World War I, and how those events would reshape labor relations in the United States. In 1912, Democrat Woodrow Wilson, the former president of Princeton University and governor of New Jersey, won a four-way race for US president, breaking what had been a decades-long Republican lock on the office. This was the first election in which New Mexico took part as a state.

On taking office, Wilson called a special session of Congress, which was also in Democratic control. Together, Wilson and the Congress enacted a series of "progressive" laws, including the Newlands Labor Act that established the Board of Mediation and Conciliation, the charge of which was to intervene in disputes between railroads and their operating workforces. The Board was a failure, but it was a harbinger of things to come. Wilson was clearly more supportive of industrial workers than his predecessors had been. The Adamson Act of 1916, for example, set a national standard of the eight-hour day for railroad employees in the operating crafts (engineers, firemen, brakemen, and conductors). And in 1917, the eight-hour standard was also extended to shopmen.[10]

War had erupted in Europe in 1914. For two and a half years, Wilson attempted to hold the United States out of that horrific conflict, but in April 1917 he finally asked for a declaration of war, and Congress consented. That precipitated the sudden and sustained dispatch of thousands of troops and the equipment and materiel to support them, mostly from East Coast ports. Getting personnel and supplies to the ports of embarkation was the job of the nation's railroads, which seemed incapable of coordinating adequately among themselves. A glut of freight cars on the East Coast clogged the port facilities; meanwhile, there was a shortage of rolling stock in much of the rest of the country. The result was a dangerous transportation crisis.

In the face of impending military disaster caused by the railroad gridlock, Wilson nationalized all the railroads on December 26, 1917, thereby turning some 2 million railroad workers into government employees.[11] In order for such a national rail system to work, the patchwork of hundreds of different work rules, pay scales, and corporate procedures maintained by the private railroads had to be replaced by a single, uniform, nationwide organizational and operational structure. Also essential was a shift in labor relations from confrontation between owners and workers to cooperation between administration and staff within a large public entity. Almost overnight, labor unions were recognized as a valuable mechanism for keeping labor relations on an even keel.

The national government became the advocate for both administration and staff, with a desire to ensure fairness to both those necessary components of a smoothly functioning railroad system. With corporate profit no longer the governing principle of railroad operations, the Wilson administration encouraged the expansion of railroad unions. For example, labor organizers began a campaign on the AT&SF in February 1918. When the company retaliated by firing some of the union shopmen, the US Railroad Administration announced that the company could not single out union members and officials. Later that year the Board of Railroad Wages and Working Conditions raised shopmen's hourly pay rate to sixty-eight cents an hour. "This award, in effect, standardized the wage levels for the nation's shopmen."[12]

As recorded in the proceedings of the 1918 convention of the Railway Employees' Department (RED) of the AFL, this was "the first time in history our Government has given full voice in the conduct of its affairs to labor."[13] Optimism within labor unions was running very high, and they were seeing results they liked.

But then the war ended, and immediately railroad stockholders and officers began agitating for the return of unfettered authority over their companies. Less than two months after the signing of the armistice ending the war, Wilson agreed that control of the railroads would be restored to their owners by January 1, 1920. Before the end of February 1920, Congress passed and President Wilson signed the Transportation Act of 1920, establishing March 1 as the date on which private owner-ship of the railroads would resume. Among many other provisions, the act also established the Railroad Labor Board, which was empowered to decide disputes over wages and salaries.[14] In what later seemed like a final act of goodwill, in July, the Board voted to raise the hourly wage for shopmen by thirteen cents.

By the time of the presidential election in November 1920, though, a national economic downturn was underway, and layoffs of shopmen began. Warren Harding, a Republican, succeeded Wilson as president in 1921, and the layoffs continued. This was part of a concerted strategy by railroad companies to break the power of unions and "re-establish the prewar relationship between the railroads and their employees."[15] The new Harding administration's sympathies were decidedly with the rail-road ownership. In June 1921 the Railroad Labor Board voted to cut all railroad workers' pay by an average of 12.5 percent.[16] The railroads now adopted new strategies to circumvent federal wage requirements: the sub-stitution of piece work for payment of hourly wages and the subcontract-ing of operation of facilities for repair of locomotives and other rolling stock. Then in July 1922 wages for shopmen were scheduled to be cut again, this time by seven cents an hour.[17]

In response, nationally on July 1, about 250,000 shopmen walked off the job, including 1,000 in Albuquerque. The following day the *Albuquerque Morning Journal* published a statement to the public from the Local Federation of Shopmen, from which the following excerpts are taken: "In behalf of the 98 percent of railroad shop employees who yesterday withdrew their services from the railroad company, we desire to give the public a brief statement of the issues involved:

([issue #] 3) The decision of the United States railroad labor board to again cut wages of the employees 56 cents per day on all mechanics, helpers and apprentices, except freight carmen, whom it is proposed to cut 72 cents per day.

. . . The third proposition, upon the wage question, is the straw that broke the camel's back, and when our friends really understand what is contemplated in the latest decision of the board, we feel sure we will have your unqualified sympathy and support. . . . In December, 1917, this class of mechanics received $5 for a day's work; under the new proposed scale of pay they would get $5.60, or twelve percent more than they received in 1917, instead of 29% more, as stated by the railroad and public members of the labor board. Why this misrepresentation, if they wished to be fair? Why also did they not cut supervisory forces in proportion, if they wanted to be fair?

. . . The proposed reduction, being the same for helpers and apprentices, will affect these classes of employees proportionately greater.

. . . We feel that this decision is a miscarriage of justice; that that portion of the transportation act requiring that consideration must be given [to] "the relation between wages and the cost of living" was disregarded; and we have therefore exercised our precious American privilege of withdrawing our services from the railroad company.[18]

Within days nationally, "a combined total of 400,000 shopmen [had] downed their tools." But "the railroads reacted swiftly and instituted a well-organized counter offensive. An army of guards was recruited to protect railroad property and the movement of strikebreakers."[19]

The AT&SF took a hard line on the subject of striking workers. As reported in the *Morning Journal* (July 7, 1922), company officials stated that "Practically all men who quit work learned their trade with the Santa Fe and . . . we can train new [men] to fill their places . . . [it will] not be a difficult task." A sample of oral histories on file at the National Hispanic Cultural Center in Albuquerque indicates that many workers in the Barelas neighborhood went back to work within weeks of the strike being called because they were fearful of losing their jobs permanently. Of the others who stayed on strike, many never returned to railroad work because they were black-listed by the company.[20]

Within a week of the beginning of the strike, scabs, or more euphemistically "replacement workers," were substituting for strikers at the Shops in Albuquerque. An accusation of working as a scab resulted in an automobile accident on July 7 at the intersection of 1st and 2nd Streets, near the western entrance to the Shops.[21] According to a long-time resident of the Barelas neighborhood in Albuquerque, the AT&SF recruited replacement workers in Mexico and brought them onto the Shop grounds in cattle cars, hidden among the cattle.[22]

On the day after the strike began, the shopmen's crafts declared the strike "100 percent perfect," although National Guard units were being put at the ready to stand behind the railway companies. This was in response to several cases in the Midwest of intimidation, both verbal and physical, of strikebreakers by union members.[23] Earlier that week, on July 6, Albuquerque machinist William Bletz, who had stayed on the job, filed a complaint against two striking pipe fitters, Gus Brito and Maurice Cowell, whom he accused of having beaten him up. Further, it was being said that massed strikers planned to storm the Albuquerque Shops, which the company had ordered stockaded.[24] Likewise, Barelas resident Jennie Bargas-García remembered that "strikers were mean to her father and called him names and pushed him" because he did not go out on strike.[25] "Whatever cause could be assigned to a particular strike, the motivation behind the decisions of individual railroaders was more complex. An individual worker had to consider his family, his chances for reemployment, and whether striking could jeopardize his ownership of his home. Moreover, he was affected by his social relations: failure to strike when most of his peers were struggling against the company could cost a man the respect and friendship of his closest associates."[26]

With Albuquerque apparently having no shortage of shopmen on the job, on July 15, carpenters and mechanics were sent from there to the shops at La Junta, Colorado, to assist with building permanent bunkhouses for replacement workers in anticipation of a possibly lengthy strike.[27] As noted earlier, the La Junta shops had been a hotbed of union activity during the 1893 shopmen's strike. This suggests that the company, in 1922, not wanting a recurrence of the earlier stubborn disruption of its business, strategically sited the replacement workers' dormitory at La Junta to intimidate especially active union members there.

In late July the AT&SF attempted to bluff its way to a systemwide settlement by asserting that with only 57 percent of its shop workforce, it was able to keep its trains running according to their usual schedules. Meanwhile, the Burlington claimed to be adding "250 to 325" new strike-breaking shop employees every day during the strike.[28] Just a week before, the Santa Fe was advertising in the *Albuquerque Morning Journal*, "Men Wanted . . . Machinists, Boilermakers, Sheet Metal Workers, Electricians, Car Men and Helpers." These were men intended to take the place of strikers.[29] Both the company and the strikers filled the newspaper with propaganda, the brotherhoods saying that they had

Figure 10.1. Scan of an ad for shopworkers placed by AT&SF on page 10 of the *Albuquerque Morning Journal*, Sunday, July 23, 1922, just over three weeks after the Shopmen's Strike began.

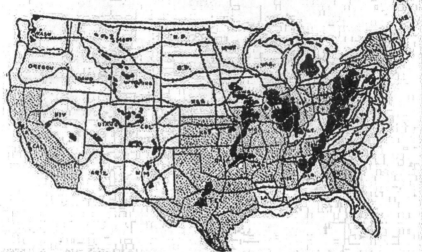

Map 10.1. Scan of national map showing railroads and coal mining areas affected by concurrent strikes by shopworkers and coal miners, top-center of page 1 of the *Albuquerque Morning Journal*, Sunday, July 23, 1922.

never had more support than they were then receiving, and AT&SF insisting that it was "now moving more freight in New Mexico than at any period in the past two years."[30] These were radically divergent portrayals of the state of affairs during the strike.

In the midst of the continuing strike, on September 23, 1922, the company launched the building of a new boiler shop at the Albuquerque Locomotive Repair Shops complex.[31] (Much more about construction of the new Shop buildings appears in chapter 11.) At this point, it is an open question as to how many of the strikers may have signed onto construction crews so that they actually kept working for AT&SF even as the shopmen's strike continued.

Strikers who took other railroad jobs within the Albuquerque Shops, elsewhere within the system or with other railroads, were commonly reviled by their fellows. Sometimes that caused the breakup of friendships and family relations, when a striker decided to seek alternative railroad employment.

Exceedingly injurious to their national collective bargaining position "were the large numbers of striking shopmen who went scabbing on railroads other than their own."[32] When strikers took work in the shops of other railroads, they made it possible, through a kind of musical chairs, for many railroads to maintain their locomotive repair schedules and, thus, to keep trains running throughout the country. As a result, the striking shopmen could never significantly disrupt the nation's daily train traffic. Had they been able to halt entirely the nationwide flow of goods and people by rail for even a week or two, public outcry probably would have brought the strike to a conclusion in short order in the shopmen's favor.

But the immediate needs of shopmen and their families for money to purchase the basics of urban life, food, clothing, and shelter produced and offered for sale by others, pushed way too many of them toward the easiest remedy. That was to move to another town and to another locomotive repair shop, where other strikers had left their jobs vacant and where they could earn wages they were used to. That could have been as easy as taking a day-long train trip to another place hundreds of miles away, where the striker wasn't known as such, and responding to an ad like those posted by the AT&SF in the *Albuquerque Morning Journal*.

On top of that, for too long the shopmen's common macho attitudes stood in the way of their seeking financial support during the strike from others in their communities. Sometimes, even accepting support from within their own families was too painful for men who saw themselves as up to any task. As Colin Davis writes, "Initially reluctant to organize women in auxiliaries, strikers in Albuquerque recognized their error: 'We made a mistake here [one of them was quoted as saying] in not earlier inviting the ladies to cooperate with us, but now that we are on the right track, we will try to make up for lost time.'"[33]

But recovery of strike momentum seemed out of reach. The strike stretched into a month, and then two, and then three. By that time, on September 15, regional railroads like the Chicago, Burlington, and Quincy (CB&Q) had begun chipping away at the national character of the shopmen's strike by entering into an "agreement embracing a new schedule of rates of pay and working rules" with only its own shop employees.[34] This became a concerted strategy of the railroads. Not only did it get shopmen back to work, but it tore apart the idea of national unions.

Then, at a September meeting of the RED of the AFL, the leadership supported the idea of separate contracts, but the rank and file fought to hold the strike together until a national settlement could be reached, arguing that any "separate agreement would, 'tear down the morale of 50% of our men on other roads'" and would disrespect the "men now in jail and the men under six feet of ground who went out for the principle of a National settlement."[35] But the union leadership held sway, and the settlement with the CB&Q was signed, confirming a rupture within the unions that would prove fatal to success of the strike.

Nearly three months into the strike, on September 23, 1922, a US District Court judge issued a nationwide temporary injunction against the strike by shopmen.[36] Thereafter, "increasingly, federal judges came to view railroads as semipublic enterprises, and thus any threat to their operation could be enjoined under the Sherman Anti-Trust Act."[37] The shopmen's strike and the strength of the shop-craft unions were effectively undermined by injunctions.

The effects of the CB&Q settlement and the national injunction are seen clearly in the numbers of railroad machinists employed nationally between June and October 1922. In June before the strike began, 55,410 machinists were on the job. With commencement of the strike on July 1, that number collapsed to only 18,070 for the month of July. In September, though, after replacement workers had been aggressively recruited by the railroads and the national temporary injunction had been put in place, the total number of machinists at work had climbed back to 38,550. And by the end of October, the national employment roster of railroad machinists, 56,680, actually exceeded the pre-strike level.[38]

Figure 10.2. Photo of button commemorating the 1922 shopmen's strike.

The local impact of the strike in Albuquerque is harder to gauge. At the time, local news sources seem to have been reluctant to make statements that might have been seen as encouraging to either labor or the company. As a result, the *Albuquerque Morning Journal* appears to have made a concerted attempt to ignore the strike, to publish news about anything but the strike when possible. So, it is in only seemingly casual, even inadvertent, printed remarks that hints about the strike's local impact emerge.

It may be that the strike's direct effect on the Albuquerque workforce was reduced because only nine months earlier shopmen had begun to return to work in the new facility after a months-long closure of the machine shop. It is unlikely that the number of employees was yet back up to what it had been before construction. That means that a pool of furloughed employees would have been locally available to take the

place of striking shopmen, thus discouraging strikers from any prolonged work stoppage.[39]

As additional local and regional contracts were signed by shop-craft
organizations across the country during the remainder of 1922 and
through 1923, dwindling numbers of tenacious local strikers continued
to hold out for what they saw as the only hope for their profession, an
equal seat at the bargaining table with the railroad owners. "By November 1923, the strikes were in serious trouble. . . . Termination of strikes
continued during late 1923 and 1924. . . . The major problem was the
inability of the shop crafts to send help [to the remaining strikers]. . . .
On many of the railroad systems the lines of strikers held tight. What
these men and their families had been unable to control, however, was
the continuing output of rolling stock from inside the railroad shops."[40]
The trains kept moving, and ultimately the American public saw the
strikes as only a painful nuisance.

By early 1925 only two regional shopmen's strikes remained active
in the United States, both on Eastern railroads. They dragged on until
1928. But already, since the failure of the strike to establish a national
shopmen's union in 1922, workers in most locomotive repair shops had
been left essentially without leverage to counter unilateral mandates by
the railroad companies with regard to wages, hours, seniority, and
working conditions. As elsewhere across the country, the shop workforce at Albuquerque had little recourse from arbitrary dictates and
decisions of the Santa Fe owners and management.

One of the impacts of the 1922 shopmen's strike in Albuquerque was
an additional influx of Hispanic workers into the Shops, many coming
from nearby farming communities, such as Belen, Los Lunas, and
Tomé, to take the place of strikers.[41] Another result was the permanent
loss of work at the Shops by many men who had gone out on strike.
Two such men were the father-in-law of Frank Archibeque of Barelas
and Rufina Salazar-Montaño's father.[42] The company's dominant concern with breaking the national shopmen's organization overrode what
had been a pervasive bias against the hiring of journeymen Hispanic
shopworkers and the promotion of Hispanic helpers and apprentices.
The net effect, fully unintended, was an increase in ethnic diversity
among the skilled workforce at the Shops.

To shed new light on this influx of Hispanics into the Shops, we analyzed data taken from *Hudspeth's Albuquerque City Directories* for 1919 and 1925, before and after the shopmen's strike of 1922. What that analysis shows is that in each of the six largest job categories, machinists, machinist apprentices, machinist helpers, boilermakers, boilermaker apprentices, and boilermaker helpers, the Hispanic share of the Shops' workforce increased from before the strike to after the strike. With the exception of boilermaker helpers, among which category Hispanos already made up 76.9 percent of the workers in 1919, in all five other job categories Hispanos saw significant increases in their share of the workforce, ranging from 10.4 percent increase for boilermaker helpers, 24.7 percent increase for machinist helpers, and 25.5 percent increase for machinists, to a whopping 71.2 percent increase for machinist apprentices. In addition, in 1925 there were a minimum of ten Spanish-surnamed boilermaker apprentices, where in 1919 there had been none. (For a broader view, see appendix 2: Ethnicity of Shopworkers in Six Job Classifications, 1919, 1925, 1943, and 1950.)

As a further indication of the convulsions within the Albuquerque Locomotive Repair Shops workforce that came about largely because of the 1922 strike, only 36 of the 198 AT&SF journeyman machinists listed in the 1919 *City Directory* (including just 7 Hispanos) are also listed in the 1925 edition of the *Directory*.[43] That translates into a huge attrition rate among journeyman machinists over that six years of 81.8 percent. In addition, only 4 men who were recorded as either machinist apprentices or machinist helpers in 1919 had moved into the ranks of journeyman machinists in 1925. The company had clearly terminated the employment of most of its technically skilled workers. As Carmen Aragón-Moya, a teacher in the Barelas area, put it, the Santa Fe "fired the workers and hired more people."[44]

Thus, despite the scanty print record of discord and disruption during the 1922 strike as registered in contemporaneous news media, it is clear that the strike utterly transformed the roster of workers at the Shops and consequently must have radically upset the lives of most Albuquerque shopmen and their families. Unemployment was followed by marginalized existence for many, often succeeded by migration away from Albuquerque.

CHAPTER 11

State of the Art

Building the New Shops, 1914–1924

W hen the original Albuquerque Locomotive Repair Shops were built in 1880 and 1881, steam locomotives were tiny compared to the behemoths that would run the rails in the 1930s and 1940s. They hauled light loads mostly over the easy terrain of plains and broad river valleys of the East and Midwest.

It was clear, when transcontinental railroads were first envisioned, that much more powerful locomotives would be required to negotiate the broken and mountainous West. Then, when railroad transportation became the norm after the Civil War, and much greater loads of passengers and freight were beginning to move by rail, the need for hugely increased locomotive power became imperative for a booming railroad industry.

The Atchison, Topeka, and Santa Fe Railroad/Railway was no passive purchaser of super locomotive power. It was a leader in the engineering, design, and manufacture of the mammoth steam locomotives that hauled a significant share of the nation's goods and passengers throughout the first half of the twentieth century. Already in 1880, under the impetus of its chief mechanical engineer, AT&SF purchased the first locomotives built to the company's own specifications from the Baldwin Locomotive Works of Philadelphia. In 1901, the railroad began using 2–10–2 locomotives (i.e., with two small front pilot wheels, ten large drive wheels, and two small trailing truck wheels)—known appropriately as the "Santa Fe Type"—built in its own Topeka shops.[1]

Each new generation (or class) of locomotives, produced in almost as rapid succession as cell phones are today, was bigger, heavier, and more powerful than its predecessor. Each new class carried more fuel and water, had a bigger firebox and bigger boiler, and consequently could pull more weight (its tractive power). As just one indication of that trend, in 1890 the average steam locomotive weighed 92,000 pounds; just ten years later the average was 120,000 pounds.[2] The accompanying table compares a selected group of five classes of AT&SF steam locomotives manufactured between 1887 and 1937. Note especially that the length of the locomotives (without tenders) approximately doubled over that fifty-year period, from 33.9 feet to 67.2 feet. An obvious consequence of that mushrooming size was that the locomotives outgrew the company's repair shops, including the shops at Albuquerque.

That trend in locomotive size came as no surprise to AT&SF because it was in large measure driven and realized by its own mechanical engineering staff. The handwriting was on the company's drafting tables, so to speak. The trajectory of locomotive size was clear from at least the early 1900s, as were the rapidly increasing volume of railroad traffic across the Southwest and the swelling number of locomotives in the

Table 11.1. Selected AT&SF Locomotives Compared by Size and Power

YEAR	CLASS	LENGTH	WHEELS	TRACTIVE FORCE	BOILER DIAMETER	FIREBOX GRATE AREA
1887	132	33.9 ft.	4–4–0	15,600 lbs.	54 in.	17.9 sq. ft.
1894	616	34.5 ft.	2–8–0	27,900 lbs.	62 in.	23.5 sq. ft.
1905	900	48.1 ft.	2–10–2	74,800 lbs.	79 in.	58.5 sq. ft.
1923	3800	60.8 ft.	2–10–2	81,500 lbs.	100 in.	88.3 sq. ft.
1937	5001	67.2 ft.	2–10–4	93,000 lbs.	104 in.	121.5 sq. ft.

Source: AT&SF Railway, "Steam Engine Diagrams and Blueprints," Atchison, Topeka and Santa Fe Collection, Railroad, Box 535, Folder 2, Item No. 221763, Kansas State Historical Society.

Figure 11.1. Crew of African American mule drivers, "Grading for the Great New Shops in Albuquerque." This photo shows part of the equipment of the Springer Transfer Company preparing the ground for the erection of additional shop facilities at Albuquerque. Although the quality of this reproduction is poor, the photograph provides a rare record of African American workers during building of the new Shops. Note the blacksmith shop in the upper right background. *Santa Fe Magazine* 9, no. 11 (October 1915): 44.

AT&SF fleet. The Santa Fe owned 839 steam locomotives in 1896. By 1905, that number had grown to 1,454, a nearly 75 percent increase, and by 1915, the fleet numbered 2,105, a jump of about another 45 percent.[3] By the middle 1910s, the combination of the steadily growing size of steam locomotives and their multiplying numbers left the Albuquerque Repair Shops painfully small to handle the necessary work of keeping AT&SF's motive power in working order, on which the company's business was utterly dependent. As the industry journal *Railway Age* stated in 1922, "It became apparent several years ago that this layout had become inadequate and consideration was given to the construction of entirely new facilities. The studies made at that time indicated that the best arrangement would be secured by constructing the new facilities on the present location. To that end, additional land was acquired in 1913."[4]

Construction of the enlarged and modernized shop buildings got underway in 1914. The original boiler and blacksmith shops and other smaller buildings south of the transfer table were demolished, and the

Figure 11.2. Albuquerque, New Mexico. Part of the Atchison, Topeka, and Santa Fe Railroad store department (storehouse) built in 1915. More than 35,000 different items were carried here. Note banks of high windows and sloped shelving with cubbies for individual items, 1943. Photograph by Jack Delano, National Register of Historic Places Registration Form, 2014, Figure 9.

land was cleared and leveled, a job performed mostly by African American laborers using mule-drawn graders and other equipment. Contractors then built a new carpenter shop and a much ballyhooed and admired storehouse (now the home of the WHEELS Museum), both built of reinforced concrete, as well as three brick lavatories and two wood-frame buildings—a coaling chute and a car repair shed.

It is worth pointing out some of the innovative features of the new storehouse. In 1915, Aubrey Wachter, then AT&SF's southwest division storekeeper, boasted that the brand new storehouse was "probably today unsurpassed by any railway supply depot in this country."[5] The one-story, reinforced concrete building still shows its original footprint:

indoor storage and office space of 409 feet x 50 feet, with additional outdoor storage on a concrete platform 800 feet x 70 feet, with deck height matching the standard deck height of railroad boxcars and flatcars. A continuous rank of large windows runs down each wall at a height of 11 feet above the floor, resulting in an extraordinarily well-lighted workspace. This attention to abundant natural lighting is a hallmark of all the new buildings in the shop complex.

Within the main indoor storage space, a supply of small and especially costly parts was kept in four long files of wooden cabinets that ran lengthwise down the longest dimension of the building. Each individual cubby or drawer was labeled with a numeric code indicating the specific part or component contained in it. Spacious aisles between the rows of cabinets permitted easy access by hand trucks and carts. Likewise, outside on the platform the locations of larger parts were indicated by painted outlines on the concrete deck. In all, the new storehouse carried more than 35,000 items.[6] Efficiency in storing and locating those items was the watchword that governed the building's design.

A variety of petroleum products—primarily assorted oils and greases—and varnishes were stored within fireproof rooms in the basement. Twenty-six self-measuring pumps dispensed these liquids to workers within an oil house portion of the storehouse. New supplies of these heavy, messy, and essential substances were delivered by a gantry crane through an opening in the oil house roof. Gasoline and other flammables were stored in another area of the complex and delivered by pipe to the storehouse.[7] To show off this new facility, the company convened a systemwide meeting of about 150 store employees at the Albuquerque Shops, September 27–29, 1915, something never before attempted by the company.[8]

AT&SF track crews also realigned the rails serving the Albuquerque Shops in keeping with what would eventually be the layout of the new complex. Issues of the *Santa Fe Magazine* for employees also recorded other renovation and expansion work undertaken within the shop complex during late 1914 and 1915: a new planing mill with a band saw for ripping and resawing boards and a planing waste exhaust system, a

Figure 11.3. "Group photo of attendees at the first meeting of the storekeepers of the entire Santa Fe System, September 27–29, 1915." Storekeepers are posed on the steps to the north entrance of the brand new storehouse at the Albuquerque Shops, currently WHEELS Museum. Photographer unknown. *Santa Fe Magazine* 9, no. 11 (October 1915): 40.

wheel shop, an additional lavatory with locker room, a tie-treating plant, and a new electrical generating plant.[9]

At that point, though, the need for coordinated nationwide transportation occasioned by the large-scale fighting of World War I interrupted business plans for the AT&SF and many other American companies. Despite the drop in railroad revenues that followed, in 1917 the AT&SF put together the funds to erect a new blacksmith shop at Albuquerque built of brick and steel. The new shop was a spacious 80 feet x 306 feet, housing "three steam hammers and heavy duty blowers to power the forges for annealing metal parts. Old driving wheels and other scrap metal were re-forged on-site."[10] The new blacksmith shop still stands and is now home to the Rail Yards Market.

As we have already seen, at the end of 1917, President Wilson ordered nationalization of the railroads, placing control of most aspects of the

Figure 11.4. North façade of the new blacksmith shop, built in 1914. Currently called The Yards, home to a seasonal growers' market, 2020. Photo by the authors.

Figure 11.5. Interior of new blacksmith shop at the Albuquerque Repair Shops, with power hammers, 1948. Note large window area on long (east and west) walls. Photograph by Barnes & Caplin. Courtesy of the Albuquerque Museum, Photo Archives, catalog number PA1980-184-907.

operation of all railroads under the United States Railroad Administration (USRA). Nationalization and the subsequent diversion of construction resources to projects directly related to the war effort interrupted further work on the new Albuquerque Shops. With the end of World War I in late 1918, however, railroad companies immediately pressed for early restoration of full private control of all aspects of the rail industry. In anticipation of that eventuality, in 1919, AT&SF entered into a contract with J. E. Nelson and Sons of Chicago to build a new machine-shop building at Albuquerque, which would cost some $3 million.[11] The USRA was abolished in April 1920 and railway companies, including AT&SF, were compensated for their depressed income during the years of the USRA's existence.

The AT&SF pushed forward immediately with its plans to enlarge and modernize the Albuquerque Shops. On November 1, 1920, Nelson and Sons began construction of the new machine shop, and building of new flue and babbit shops had started two months earlier.[12] "The decade from 1914 to 1924, when . . . new Albuquerque shops were constructed [to replace the 1880s buildings], coincided with the greatest period of innovation in the history of industrial design and building technology."[13] As a result, the machine-shop, boiler-shop, and flue-shop buildings, akin to the contemporaneous Ford Motor Company's "Glass Plant" at its River Rouge, Michigan, complex, represented the state of American industrial architecture, featuring glass-curtain walls and skylights, supported by a stunning steel framework and enclosing a huge, airy, open workspace. The interior organization of the various Albuquerque Shop buildings was also certainly influenced by studies such as that carried out by Harrington Emerson from 1904 to 1907 at AT&SF's Topeka Shops. His principles of efficiency became a guide for many other shops as well.[14]

The new buildings marked a wholesale departure from earlier norms of American industrial architecture, which relied on stone masonry walls and wood-frame roof structures. The new steel girder framework permitted both quicker erection of the buildings and the incorporation within them of much wider roof spans that enclosed huge work floors. When completed, the three-and-a-half-acre machine shop, for instance,

Figure 11.6. New machine shop under construction, showing steel girder framework and peaked framing for one of many skylights, 1922. Photographer unknown. Courtesy of Center for Southwest Research, University Libraries, University of New Mexico; Albuquerque Construction Sites Album, 1919-1923, PICT 2002-013-0002c.

Figure 11.7. (*below*) Interior of the erecting bay of the machine shop at the Albuquerque Locomotive Repair Shops, showing north steel and glass curtain wall, 2014. Note inspection pit in center foreground. Photograph by Petra Morris, National Register of Historic Places Registration Form, 2014, number 0005 of 0052.

could accommodate as many as twenty-six steam locomotives simulta-neously, a larger capacity than the AT&SF's principal shops in Topeka, Kansas.[15]

By the end of August 1921, glaziers were already installing the thou-sands of 14 x 20 inch panes of special ribbed glass manufactured by the Streator Glass Company.[16] These formed the long northern and south-ern walls of the machine shop and were its "most noteworthy feature," one that is still the most arresting aspect of the building.[17] Essential to the operation of the machine shop, which had to be capable of moving locomotive parts weighing many tons, was the 256-ton overhead travel-ing crane, as well as five other, smaller-capacity overhead cranes. The cranes' rails and supporting armatures were part of the building's steel scaffolding. The floor, too, was innovative. It looked like it was brick, but, in fact, was made of creosoted wood blocks laid on a concrete slab. This floor system was more comfortable to stand and walk on than concrete alone, helped suppress the clatter of dropped parts and tools, and reduced the likelihood of sparks from falling metal objects. Each locomotive bay was defined and provided ingress and egress by a short stub of track and a central, longitudinal, concrete service pit, giving ready access to each locomotive's underside.[18]

The floor plan of the machine-shop building, huge for its day, was divided into four parallel bays, each with a rectangular footprint with the long axis oriented east-west. The largest and most northerly of the four spaces was the erecting bay, where the major work of disassem-bling and reassembling steam locomotives took place. The dimensions of its floor space were 600 feet x 90 feet, and it had a usable vertical space of 57 feet. This is where the twenty-six concrete service pits—each 63 feet long—and the 250-ton crane were located, as well as two 15-ton overhead traveling cranes.

At any given time, the erecting bay would house multiple locomotives in various stages of overhaul, plus equipment necessary for breaking down and reassembling those complex machines, including "fifteen engine lathes; a car-wheel lathe; a double-head car axle lathe; two verti-cal and one horizontal turret lathes; one 100-inch boring mill and four smaller mills; five radial drill presses; one double-head and five

Girders for 250-Ton crane. Albuquerque Machine Shop. 2-10-22.

Figure 11.8. Arrival of mammoth girders to hold 256-ton traveling crane in the erecting bay of the machine shop at Albuquerque, 1922. Photographer unknown. Courtesy of Center for Southwest Research, University Libraries, University of New Mexico; Albuquerque Construction Sites Album, 1919-1923, PICT 2002-013-0041e.

Figure 11.9. Floor area of new machine shop at Albuquerque, with concrete-lined inspection pits recently finished. Old machine shop and new blacksmith shop in background, 1922. Photographer unknown. Courtesy of Center for Southwest Research, University Libraries, University of New Mexico; Albuquerque Construction Sites Album, 1919-1923, PICT 2002-013-0007a.

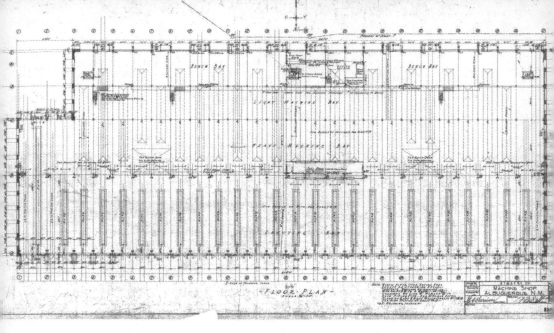

Figure 11.10. Floor plan of the new machine shop at Albuquerque, with the various bays labeled, 1932. E. A. Harrison, architect for AT&SF, National Register of Historic Places Registration Form, 2014, Figure 14.

single-head shapers; three slotters, one piston-rod and one guide grinder; three double-head dry grinders and two single wet-tool grinders; and other smaller machine tools."[19] In addition, there were multiple hand trucks, carts, jacks, ladders, and assorted wrenches, punches, hammers, chisels, snips, saws, files, clamps, bearing pullers, welding torches, and other hand tools. The net affect was that, despite the cavernous size of the space, it often felt like the machinists, apprentices, helpers, welders, sheet metal workers, pipefitters, electricians, and inspectors had barely enough room to move around.

Within the machine shop, immediately south of the erecting bay was the heavy machinery bay, a space 600 feet long and 65 feet wide. Here is where locomotive wheels could be resurfaced and balanced and where steam cylinders and their corresponding pistons, as well as the various drive rods and tie rods could be reconditioned. The heavy machinery bay was also equipped with two 15-ton overhead traveling cranes and numerous jib cranes affixed to structural columns, as well as an array of machine and hand tools paralleling those in use in the

Figure 11.11. Wheels in heavy machine bay of the Albuquerque machine shop, 1943. Note the haziness–fine particles of metal and grinding dust produced by such operations as welding–that was typical of the air in the machine shop. Photograph by Jack Delano. Courtesy of the Library of Congress, catalog number LC fsa 8d27208.

erecting bay. With a ceiling height only 60 percent of that in the erecting bay, the heavy machinery bay still felt spacious even when the floor area was thick with massive drive wheels and rods. Part of this bay was occupied by a locked 50 feet x 20 feet hand-tool room, from which workers had to check out the tools they needed.

The next subdivision of the machine shop to the south of the heavy machinery bay was the smaller-still light machinery bay, at 540 feet long and 40 feet wide, with a 20-foot ceiling height. Here parts such as light rods and levers, pilots (cowcatchers), couplings, brake parts, cable guides, whistles, bells, headlights, automatic oilers, and other smaller locomotive parts were cleaned and repaired. Again, this bay boasted a 5-ton monorail crane and many jib cranes, and it was supplied with grinders, buffers, compressed air hoses, and flexible wire brushes, together with pliers, light hammers, screw drivers, drills, files, and myriad other small machine and hand tools.

The fourth and final bay within the machine shop was the bench bay, the same size as the light machinery bay. But more of its space was occupied by small storage rooms, a blueprint room, and the general foreman's office. The repair work done here focused on small pieces of equipment, such things as gears and gauges, which often had to be held steadily in place on a flat surface or required intense light or magnification to permit manipulation of extremely small components such as washers, springs, escapements, clutches, and so on.

Even in the face of the rising possibility of a shopmen's strike, on June 9, 1922, the company announced the award of a contract to C. A. Fellows Construction Company of Los Angeles to tear down the old machine shop to make way for a new boiler shop. Fellows also built all the remaining buildings of the new shop complex, except the machine shop.[20] Less than a month later, on July 1, a thousand shopmen went on strike in Albuquerque, but that hardly seemed to put a pause on the construction work. Demolition of the old machine shop and preparation of the site were nearly complete by September 20, when it was reported that "a few of the men who went on strike with the shopmen's union on July 1 are applying for employment in the Santa Fe shops. . . . However, there has not been a large or general application for re-employment. [An] official said that the shops and roundhouse now have ninety percent of a normal [work]force."[21]

In passing, the *Morning Journal* also reported that a dining hall "for the accommodation of the men who reside on the premises," referring to strike-breaking replacement workers, had "recently been erected," which would "be used for shop purposes when strike conditions are out of the way."[22] In Albuquerque, as in much of the rest of the nation, the strike would effectively be over very soon. The signing of the CB&Q contract and the issuance of a nationwide injunction against the strike (discussed earlier) both occurred that same month, signaling the futility of continuing the strike. That did not end the bitterness between the AT&SF and many of its former and retained or rehired shopmen, but work on the new Albuquerque Shops now went barreling ahead.

The new boiler shop, a smaller look-alike of the new machine shop, was sited parallel to and just north across the transfer table on the site of the old machine shop. Like its larger mate, the boiler shop had long glass-curtain walls on its north and south sides, and a wide, high-ceilinged workspace bounded by a framework of steel girders. The boiler shop was erected during 1923, completing the core of the new Locomotive Repair Shops. The company was active in the design of all the new buildings: "The Albuquerque shops' steel buildings were designed by an in-house AT&SF design team led by C. F. W. Felt, Chief Engineer; E. A. Harrison, Chief Architect; and A. F. Robinson, the company's Bridge Engineer."[23]

The completed Locomotive Shops occupied the northern portion of a nearly level, roughly twenty-seven-acre site south of modern downtown Albuquerque between 2nd Street and the original main line of the AT&SF tracks and between Atlantic Avenue and Cromwell Avenue SW. When the repair and overhaul facility was fully enlarged and renovated by 1925, the shop complex was comprised of more than twenty-five distinct buildings, each serving a different purpose. The major buildings were the machine shop, boiler shop, blacksmith shop, tender repair shop, flue shop, cab paint shop, babbit shop, sheet metal house, wheel shop, storehouse, roundhouse and turntable, power plant, car repair shop, carpenter shop, planing mill, five lavatory buildings, locker rooms, fire

Figure 11.12. Tender of locomotive #3874 exiting the machine shop onto the transfer table, 1943. Operator's cab half in shade. See figure 11.13 below. Photograph by Jack Delano. Courtesy of the Library of Congress, catalog number LC- fsa 8d15537.

Figure 11.13. Operator's cab of the transfer table near western end of the track, 2020. Boiler shop to the left, machine shop to the right. Photo by the authors.

station, administration building, and coal and sand towers. In addition, there was a tie treatment plant, as well as lime, cement, paint, gasoline, and arch brick storage buildings, and several sheds and smaller buildings. The entirety of the complex was interlaced by tracks and concrete fire access roads, as well as a system of eleven fire alarm boxes, including ones at the machine shop, in the middle yards, and in the lower yards.[24] And all of the Shop buildings were linked together by gas, telephone, and electric lines, as well as steam and cold water pipes, and a sewer system. There was also a 600-foot-long transfer table located between the machine shop and the boiler shop. Its purpose was to move locomotives on an east-west axis, at a right angle to the main and subsidiary rail lines, into and out of the machine shop bays.

When completed in 1925, the Albuquerque Locomotive Repair Shops represented the state of the art in American industrial design. The Shops were also supplied with the most up-to-date machining and other technical equipment in the railroad industry. So much were the Albuquerque Shops seen as the model for modern steam locomotive

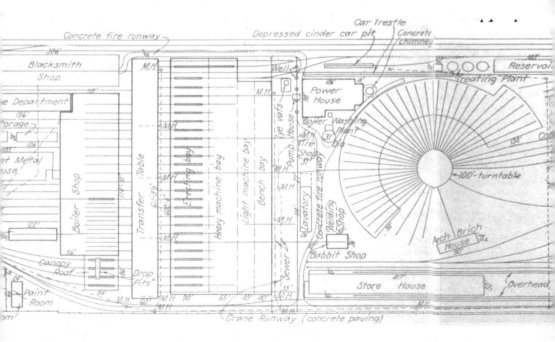

Figure 11.14. Cropped version of a site plan showing the layout of new [1920s] and old [1880s] buildings at the Albuquerque Locomotive Repair Shops complex. New buildings shown in heavy lines, "Santa Fe Completes Modern Shops at Albuquerque," *Railway Age* 73, no. 6 (August 5, 1922): 238–39.

Figure 11.15. Machine shop at AT&SF's repair shops in Cleburne, Texas, 1935–1945. Photographer unknown. Courtesy of the Kansas State Historical Society, DaRT ID: 51380.

repair and overhaul facilities that, with only slight modifications, the plan of the whole complex was duplicated at Cleburne, Texas, after the disastrous fire that destroyed most of the shops there in 1922.[25]

As with the original arrival of the railroad in Albuquerque, building the new shops brought another influx of outsiders with special skills into town. Because both contractors hired to put up the new shops were based outside New Mexico, they brought to the job experienced employees and subcontractors. Thomas La Rue, for example, was a structural steel worker who came to work on the new boiler shop. He had previously lived in El Paso, from which he brought his wife and daughter for the time he was employed in work on the shop by V. E. Ware, also of El Paso. The little family rented a furnished room on the 400 block of Lead

Avenue SW. We know about his employment and his family because of his suicide just after Christmas 1922. La Rue was reported to have suffered a "breakdown, as a result of worry over the work he was doing on the new Santa Fe boiler shops."[26] Four other steelworkers are shown with part of their names on a 1921 photograph of the ongoing erection of the steel framework of the machine shop: Sully, Charley, Mesey, and Barr. Other, apparently managerial, personnel appear or are referred to in other photos from the same album: Howard Weir, Bill, Fred, Van Ness, Tom [Thomas E.? assistant foreman of the Shops], Sadler, and Conklin, as well as a steam shovel operator named Roberts. Yet another

Figure 11.16. On-site office of J.E. Nelson & Sons, the general contractor for building the "new" Albuquerque Shops during the early 1920s, 1922. View from the east, with new roundhouse in background. Photographer unknown. Courtesy of the Center for Southwest Research, University Libraries, University of New Mexico; Albuquerque Construction Sites Album, 1919–1923, PICT 2002-013-0013a.

photo in that album shows the on-site office of the general contractor J. E. Nelson & Sons.

Work on the new machine shop began in October 1920, and the shop was fully operational by June 1922. Building this huge state-of-the-art industrial facility took just twenty months from start to finish. That was one of the anticipated advantages to the company inherent in steel-and-glass-curtain-wall construction. It is worth noting that its completion came just weeks before the beginning of the 1922 Shopmen's Strike.

CHAPTER 12

The Heyday of the Shops

1925-1950

For the city of Albuquerque and the Locomotive Repair Shops, as for America in general and indeed most of the developed world, the period between 1925 and 1950 was one of extreme political and economic swings. It saw, for instance, both the lowest and the highest employment rates ever at the Albuquerque Locomotive Repair Shops, which reflected the national economic roller coaster.

As seen in chapter 10, in the aftermath of the 1922 Shopmen's Strike, there was a wholesale replacement of staff at the Albuquerque Locomotive Repair Shops. The result was a younger, more Hispanic workforce, although foremen and managers remained uniformly non-Hispanic. Both the company and employees evidently made an effort to get permanently beyond the bitterness engendered by the strike. By 1924, the company boasted of a smoothly operating shop facility at Albuquerque, one that ran with more productivity than ever. That was attributed at the time in large measure to more harmonious relations between the shopmen and company management. A 1924 article in the *Railway Mechanical Engineer* gave some of the credit for this turn of events to the company having initiated monthly council meetings to discuss matters of shop welfare. Employees were encouraged to attend and discuss ways to improve working conditions and efficiency. According to the article, there were also fifteen-minute noon meetings held three times a week to discuss specific topics, such as safety, wage rates, and methods of eliminating waste in the workplace.[1]

This was part of a national shift by railroads, exemplified in an article written by the vice president of the Pennsylvania Railroad, E. T. Whiter, in the December 1923 issue of *Railway Mechanical Engineer.* "Railroad management realizes that contented, healthy employees, mentally and physically, are one of the greatest assets a railroad can have," he wrote.[2]

A lasting improvement in labor relations at the Albuquerque Locomotive Repair Shops is attested by the career of boilermaker Thomas C. Cordova. A member of a family from Cabezón, New Mexico, in the valley of the Río Puerco of the East, Thomas joined the employ of AT&SF at the Albuquerque Shops in 1927. He was hired as a boilermaker apprentice, became a journeyman boilermaker, and ended his career in 1976 as a welder. From the union's founding until even beyond Cordova's retirement and up until he died in 1981, he was secretary-treasurer of the International Brotherhood of Boilermakers, Local #76, in Albuquerque. He is remembered for enthusiastically repeating that he was "the luckiest man in the world." The Shops were "the best place to work," and he had the best job at the best company there was. He was immensely proud of the benefits the Brotherhood had brought to workers at the Shops. Every month he attended the meeting of Local #76 at the VFW Hall, located in 1950 at 416 N. 2nd. For decades Cordova collected dues and kept the books for the union, in addition to supporting efforts to improve the pay, benefits, and general welfare of all Shop employees.[3]

Thomas Cordova began his nearly fifty-year association with the Shops during a period of extreme national optimism, but the economy was also heading toward the worst financial collapse in the country's history. As John Stepek recently and succinctly writes, "From 1923, America was on a roll." By the end of the decade, about 60 percent of American families owned a car, most had electricity, and 40 percent or so had a telephone. Prosperity seemed to be everywhere, paid for to a significant extent by personal debt. The stock market was soaring. It looked to some as though the United States had stumbled into a permanently rising standard of living.[4]

For many decades Albuquerque's well-being was tethered to the fluctuating success of the Santa Fe Railroad, specifically to the volume of work at the Repair Shops and the size of its workforce. Thus, in the 1920s, Albuquerque shared in the good times and exhibited many of the same trends as the country at large, while railroads were booming.

In 1910, 470 motor vehicles were registered in New Mexico. Just ten years later the number was more than 17,000, and by 1930 it was about 84,000. That represented a quintupling of motor vehicles on New Mexico's roads during the 1920s.[5] That switch from horse transport to reliance of cars and trucks came about in tandem with improvement of arterial streets and roads. "Albuquerque in the decade of the 1920s invested in paving the city's major east-west streets—Central, Grand, and Coal, and its north-south thoroughfares—2nd, 4th, Broadway, and Edith. Between 1926 and 1929, twenty miles of streets were paved." A state highway system came into existence after the US Federal Aid Road Act of 1916 became law. In 1926, what had been New Mexico Highway 1 became US Highway 85, crossing the state from north to south and passing through Albuquerque. At the same time, the main east-west road in the state became part of US Highway 66, again passing through Albuquerque.[6] More all-weather roads encouraged more car ownership, and more car ownership drove a demand for more engineered, hard-surfaced roads.

In 1928 Albuquerque's trolley company went out of business, and much of its trackage was torn up. In its place, the privately owned Albuquerque Bus Company opened for business. With an ever-widening web of routes, the gasoline powered buses shuttled Shop employees as well as other workers, shoppers, and students all over Albuquerque for nearly forty years until the city purchased the system. The City has operated and expanded it ever since.[7] Freed from the constraints of the fixed iron rails of the trolley system, bus routes were much more flexible, which made it increasingly easy for shopworkers to live in housing developments on Albuquerque's expanding perimeter rather than only in the original residential core of the Barelas and South Broadway neighborhoods.

On January 1, 1929, the manager of the Albuquerque Bus Company reported that after one year of operation, bus service was being provided in eight buses, up from the original five. That number was

Figure 12.1. Two Albuquerque buses outside the Albuquerque city bus garage, 1956. Bus on the left dates from the 1950s; the one on the right from 1928. Photographer unknown. Courtesy of the Albuquerque Museum, Photo Archives, Catalog No. PA1982.180.066.

expected to grow again in 1929. In October of that year the bus company's schedule showed six distinct routes in operation: East Central, West Central, East Silver Avenue, Sawmill, North 4th Street, and South Edith. The last of those routes served the Locomotive Repair Shops, disembarking passengers eight very short blocks east of the Rail Yards' east gate.

Two years later, on February 2, 1931, a full-page ad in the *Albuquerque Journal* trumpeted the advantages of living in the Monte Vista Subdivision, "Albuquerque's fastest-growing subdivision," east of the University. One of those advantages was "the new Monte Vista public school, Albuquerque's most beautiful public building, [which] was opened to the public yesterday and today 217 lucky children are

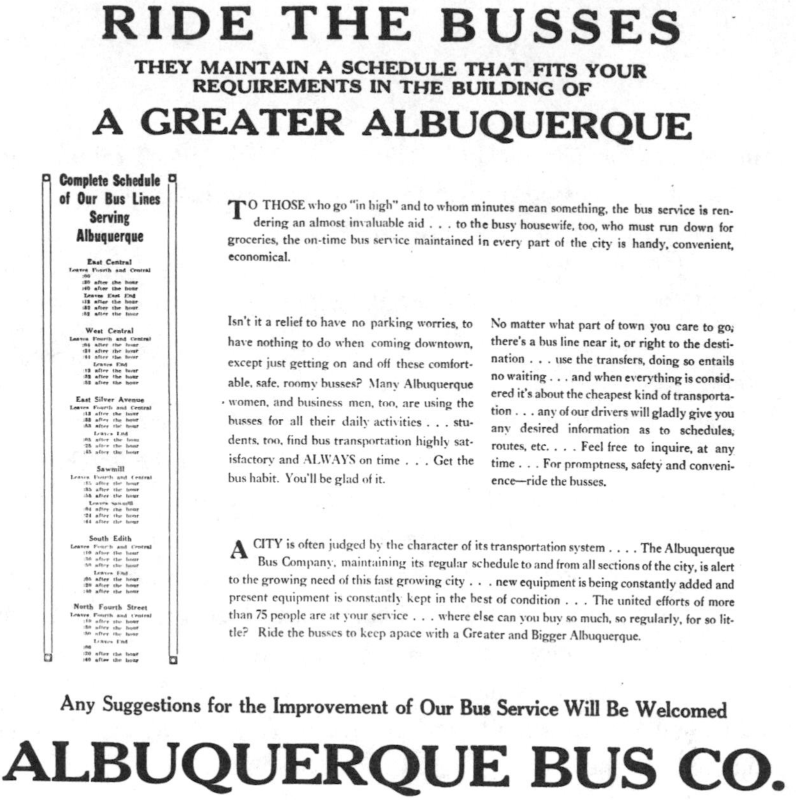

Figure 12.2. Schedule of the Albuquerque Bus Company, October 1929. Of the six bus routes, it was the South Edith Route that most directly served the Albuquerque Locomotive Repair Shops. *Albuquerque Journal*, October 7, 1929, p. 5.

attending classes there." Furthermore, the ad boasted, "The Albuquerque Bus Company runs a regular schedule of their fast, luxurious buses through Monte Vista." Less than two years after that, the newspaper announced the addition of one more bus trip per day to the new Veterans' Hospital.[8]

Driven principally by employment ramping up at the enlarged Locomotive Repair Shops during the last six years of the 1920s, the original Albuquerque town site saw explosive in-fill residential building. In the Barelas neighborhood, between 1900 and 1920, the number of houses had already jumped 330 percent to 615. Likewise, "the number of houses in the various Highlands additions, east of the railroad tracks had increased to more than 1,000." That nearly frenetic building pace continued. The newer Raynolds Addition—south of Central and west of 8th Street—witnessed similar spectacular growth, increasing during the 1920s from just 50 houses to more than 240 by the early 1930s. That meant an average of more than one new house per month in that one neighborhood alone. The Raynolds Addition was far from unusual; between 1900 and 1940 in excess of 300 new subdivisions were laid out in Albuquerque.[9]

The largest subdivision undertaken in the 1920s, at 156 acres, was the Huning Castle Addition between 15th Street and the Rio Grande, south of Central Avenue. That stretch of land had been a farm owned by Franz Huning, one of the original site developers of New Town, surrounding the Locomotive Repair Shops in the 1880s. But in 1928 the land was bought by A. R. Hebenstreit and William Keleher, son of David Keleher, one of the early Shop employees. An important component of their new development was the establishment of the Albuquerque Country Club golf course to replace the earlier sand-green course in the arroyo where Lomas Boulevard is today, near the University of New Mexico main campus.[10] The new course was of particular interest to the instructor of apprentices at the Repair Shops, Joseph Swillum, who was an avid golfer his whole adult life. As pointed out in chapter 9, he had a work schedule that permitted him to get in a round of golf nearly every Wednesday during the warmer months.[11]

Albuquerque's exuberant residential growth naturally had its counterpart in building in the city's commercial heart, centered on 3rd and 4th Streets and Central Avenue, as well as adjacent streets. In 1922, for example, Albuquerque gained its first high-rise building, the nine-story First National Bank at 3rd and Central. That was followed one year later by the Franciscan Hotel at 6th and Central and a year after that by the six-story Sunshine Building at 2nd and Central. As William Dodge notes, "The *piece de resistance* of downtown architecture in the 1920s was the KiMo Theater," which opened in 1927 at Central and 5th. "Neighborhood businesses also flourished helped in large part by the dramatic seventy-five percent increase in the city's population during the decade."[12]

Figure 12.3. Albuquerque Country Club clubhouse under construction, 1920s. Photo by Joseph Swillum, Apprentice Instructor at the Albuquerque Locomotive Repair Shops. Courtesy of Mary Jeannette Swillum Koerschner.

New Mexico's embrace of the automobile spurred the opening in the early 1920s of dealerships for car companies, including the Buick Automobile Company, the Simms Motor Company, and the Galles Motor Company, all on Central Avenue.[13] Among purchasers of automobiles during the early 1920s were, of course, employees at the Locomotive Repair Shops. Those included—to mention only a few from 1920— Assistant Shop Foreman C. L. Bernstdon, who in January bought an Oakland Six; Material Supervisor W. E. Blood, who purchased a Mitchell in October; and instructor Joseph Swillum, who became the owner of a Nash automobile in January.[14] These car owners were all among the better paid members of the Shop workforce, but many employees bought cars during the 1920s. The mechanical know-how and skill of many shopmen made cars intriguing for them and also permitted them to keep the vehicles running at far less trouble and cost than for the average car buyer.

Increasing reliance of Albuquerqueans of all walks on automobiles in the 1920s and the concomitant proliferation of car repair shops and dealerships resulted in competition among those businesses and the AT&SF Shops in hiring machinists. Over the years quite a number of AT&SF machinists quit their jobs to move to positions as automobile mechanics, among them A. F. Blank, who was reported in January 1920 to have "resigned [from the Shops] and [to then be] working at his former trade as an automobile mechanic in an uptown garage."[15] Michael Keleher remembered a former AT&SF machinist named Joe García who left the Shops and got a job at Galles Chevrolet, where he subsequently worked for about forty years.[16] Loss of experienced, skilled shopmen to the automobile industry irked the railway because it seemed as though automobile businesses were poaching AT&SF's apprentices. In effect, that meant that the Santa Fe Railway provided technical training and education to employees of other businesses. Such loss of experienced machinists from the Locomotive Repair Shops certainly added to the company's concern to try to keep its employees contented. It may help explain in part the AT&SF's decision in 1926 to "expand their medical facilities [at Albuquerque] for employees (now numbering well over 1,000)." The company constructed a three-story hospital complex at Central Avenue and Elm Street, later known as Memorial Hospital—today the Hotel Parq Central.[17] Free medical treatment for AT&SF employees and retirees at such a modern facility was a considerable perk that came with employment by the Santa Fe.

The Locomotive Repair Shops were humming and Albuquerque was growing by leaps and bounds when the stock market experienced a catastrophic slide in October 1929. Nationwide deflation set in during 1931, dragging down national economic activity, including in the railroad industry. The US Bureau of the Census recorded that the total value of all finished commodities in five categories, perishable, semidurable, consumer durable, producer durable, and construction materials, plunged from $37.8 billion in 1929 to less than half that amount, $17.7 billion, in 1932. Likewise, the number of railroad passenger trips plummeted, from 786.4 million in 1929 to only 434.8 million in 1933.[18] Considering the US economy as a whole, "by the time that F[ranklin] D. R[oosevelt] was inaugurated president on March 4, 1933, the

Albuquerque, New Mexico   15964

Figure 12.4. New AT&SF Association Hospital, Albuquerque, NM, 1925? Photographer unknown. Courtesy of Palace of the Governors, Photo Archives., (NMHM/DCA), negative number 163617.

banking system had collapsed, nearly 25% of the labor force was unemployed, and prices and productivity had fallen to 1/3 of their 1929 levels."[19]

In the early 1930s, with decidedly fewer passengers and less freight being transported by rail, fewer locomotive repairs and overhauls were necessary. In response, the Santa Fe reduced its workforce dramatically. At the Albuquerque Locomotive Repair Shops, "the low point in the Shops' operation was in 1933 when only 300 men were employed 3 days per week."[20] What had been a constant hive of activity and noise was now often nearly silent and empty of workers. Figures from the Bureau of the Census show, however, that the total value of all finished commodities in the United States in the same five categories mentioned above began to rise significantly again in 1933 all the way through the 1940s and beyond, with only a relatively small setback in 1938. For example, the annual commodity total for 1934 stood at $23.2 billion, up 30.7% from the total for the low point in 1932.[21]

For railroads, including the AT&SF, the decade of the 1930s "was colored by two major influences: the prolonged period of business prostration, with the attendant collapse of business and agriculture; and the effective competition with other forms of transportation," especially long-haul trucks and passenger buses.[22]

Additionally, as early as the 1870s and 1880s, several inventors in Europe had begun experimenting with internal combustion engines in which fuel was ignited by compression. By the late 1890s, Rudolf Diesel began licensing his design of a highly efficient engine of this type, which ever since has borne his name. The diesel engine first became popular in stationary power plants and ocean-going vessels especially because of its reliability and durability. Diesel-electric-powered locomotives ran many times longer between shop visits for maintenance and repair. There were also fewer stops for refueling during each trip, and of course no stops for water. Traveling from Chicago to Los Angeles behind a diesel locomotive automatically trimmed hours off the travel schedule when compared to steam.

With ridership on the Santa Fe down significantly during the 1930s, as well as on most other railroads, the company sought to entice passengers with faster travel. Seeking to increase the speed of transportation of both goods and passengers and to reduce the cost of human labor involved in the maintenance of steam locomotives, the AT&SF turned increasingly to diesel power.

Census data show that the first commercial diesel locomotive used by one of the major (Class 1) railroads in the United States was reported in 1925. It was strongly held by AT&SF management in the 1930s that switching to diesel locomotives would mean huge savings on maintenance and repair costs and drastically reduced outlay for purchase of locomotives. The first diesel locomotive on the Santa Fe line began service in 1935, at almost the midpoint of the Depression. The next year, in May 1936, a new streamlined, diesel-powered Santa Fe train, the Super Chief, began making regular runs from Chicago to Los Angeles. The Super Chief averaged about fifty-seven miles an hour, and in 1936 ran one 200-mile stretch of track averaging more than eighty-seven miles an hour! James Marshall writes, "The fastest steam locomotive hauling a freight train between Chicago and Los Angeles took nine engines and 35 stops for fuel and water, whereas in 1938, a

Figure 12.5. AT&SF Super Chief diesel locomotive, 1935. Photographer unknown.

diesel-powered train made the trip using only one engine and five fuel stops, cutting four to six hours of travel time.[23]

Such remarkable speeds were attributable not only to the switch to diesel power. The Santa Fe also substituted light-weight, stainless steel passenger coaches for wooden and cast iron ones, cutting the weight of its passenger trains about in half. By the end of the 1930s, AT&SF owned the largest fleet of light-weight, streamlined, diesel-powered trains of any American line.[24]

While these developments helped AT&SF weather the prolonged economic crisis, their effects were ominous for shopmen at Albuquerque and other major steam locomotive repair shops. But the brewing employment catastrophe for them was postponed, first by the continuing Depression and then by World War II. During the 1930s, AT&SF added more diesels than steam locomotives to its fleet, but purchase of new locomotives of both types remained far below pre-Crash levels. As a result, the number of diesel locomotives remained at "demonstration" levels through World War II. That was, though, only because locomotive manufacturing capabilities were diverted to production of war materiel and transport of troops and equipment.

Even before entry of the United States into the War, the national economy began to accelerate out of what had been a slow, modest recovery from the depths of Depression. By 1940, war in Europe and Asia was already driving a rapid rise in the volume of US railroad traffic. Also "the entry of the United States into the war late in 1941 brought record volumes of passengers and freight. Formerly the struggle was to get traffic; now the struggle was to handle traffic."[25] "During the war, the railroads carried 90 percent of all military freight and 98 percent of all troop movement."[26] That brought with it an equally sudden and urgent ramping up of work at locomotive repair shops across the nation. Recruitment of more shopmen in Albuquerque, especially any with previous experience or training, became a pressing priority. So much so, that "women were hired at the shops for first time."[27]

Among those hired on at the Albuquerque Locomotive Repair Shops as the workforce was expanding rapidly in 1940 was Eloy Gutiérrez, a resident of the Barelas neighborhood, then sixteen years old. Both his father and grandfather had worked at the Shops, as did his brothers Fred, Tony, and Frank and at least two uncles. As pointed out earlier, the employment of multiple family members was not unusual at the Shops. Eloy worked summers and Christmas vacations as an apprentice sheet metal worker while he was still attending Albuquerque High School.

Eloy's pay rate at the Shops was originally thirty-four cents an hour. Like all apprentices, Eloy took classes offered at the Shops by apprentice instructors in preparation for becoming a journeyman machinist. Decades later, he remembered specifically receiving instruction in the making and reading of blueprints. With the War raging when he graduated from high school, Eloy was tempted to enlist, but he was too young and continued to work in the Shops instead. Finally, though, he did enlist, still short of his twentieth birthday. He served with the Aviation Engineers, helping to build the new airstrip on the recently recaptured Pacific island of Guam. With the end of the War, Eloy returned home to Albuquerque and again took work at the Locomotive Repair Shops. Going over his experiences much later, he recalled his wife Cecelia (Cinocca) making his favorite lunch every day, an egg salad sandwich, which he usually ate in the brick layout building within the Shops complex, even though he was only three blocks from home.

Eloy remarked that during his years of part-time work at the Shops he knew an African American man who was employed pouring brass for bearings at the babbit shop. Although the Santa Fe employed many African Americans throughout the system, they were usually confined to menial and unskilled jobs, as well as passenger car conductors and porters, so the skilled babbit man stood out in Gutiérrez's memory. He also remembered one Native American working at the Shops as an electrician.

Although a number of his relatives made careers of working at the Shops, with his brother Frank eventually becoming equipment supervisor for the entire Santa Fe Railway, Eloy decided instead to enroll at the University of New Mexico and followed that with four years at Washington University in St. Louis, where he earned a degree in dentistry. In 1954 he opened a dental office in Albuquerque where he became a highly respected professional and passed on his involvement in dentistry to his children, retiring himself in 1988.[28]

Prior to US entry into the War in December 1941, the Army Air Corps established Kirtland Army Airfield on Albuquerque's Southeast Mesa as a training facility for bomber pilots and bombardiers. As a result, "many businesses in Albuquerque were awarded military contracts. The AT&SF was in the forefront of moving these manufactured goods to their destinations."[29] Two military bases in Albuquerque, Kirtland and Sandia Army Base, relied on rail service to transport men, supplies, and materiel. The number of people on payroll at the Locomotive Repair Shops continued to grow. In 1943, while Eloy Gutiérrez was getting ready to ship out to the Pacific, the AT&SF workforce in Albuquerque reached 1,787, a more than fourfold increase since ten years earlier. At $3.5 million a year, the Santa Fe was meeting "the largest single payroll in the community other than that of the Federal

Figure 12.6. Aerial photo of Kirtland Army Airbase, Albuquerque, NM, April 1942. View to the east. Note bomber pilot and bombardier training school and VA Hospital. Photo by unknown airman or Air Corps employee.

Government." The Locomotive Repair Shops were "busy day and night," running nine-hour shifts and completely overhauling about "forty-one locomotives a month."[30]

During the War, the AT&SF and all other railroads had to compete with the US military and military contractors to keep their shops at full staffing. Even in the immediate aftermath of the War, there was no let-up in the need to ship supplies and transport people by rail, now to deal with rebuilding a war-ravaged world and to satisfy long pent-up domestic demands for family housing and retail space. Once again, that offered opportunities to many Albuquerqueans.

One of those to seize that opportunity was Mike Baca, born at Five Points in Albuquerque's South Valley, who began working at the Shops in 1945, at age sixteen, as an apprentice machinist. As with Eloy Gutiérrez, Mike's father and grandfather had also worked as machinists at the Shops. Machinist apprentices, including Mike, were now making sixty-three cents an hour, nearly double what Gutiérrez had been paid just four years earlier. At the same time, Mike's father, journeyman machinist Refugio Baca, who helped Mike get his job, was paid ninety-six cents an hour.

There were a number of parallels between the Baca and Gutiérrez stories, in addition to the number of their relatives who worked at the

Shops. Both Mike and Eloy were good students, who took quickly to the apprentice evening classes in mechanical drawing and excelled in their shop duties. In Mike's case, that meant running a precision boring mill to fabricate brass bearings. Neither Mike nor Eloy ended up making a career with the railroad; instead, both earned college degrees and joined the growing ranks of Hispanic professionals in Albuquerque.

Mike's father supplemented his machinist's pay by selling lots in a small housing development he established on Baca Street in the Barelas neighborhood. He also used his free railroad pass from AT&SF to travel frequently to El Paso to visit relatives who had moved there during the 1920s Shopmen's Strike. Refugio had a fraught relationship with a particular supervisor at the Shops. That supervisor one day ordered him to pick up a 300-pound locomotive part under pain of dismissal if he did not. Refugio attempted to lift that weight, suffering a hernia in the process. For years afterward he suffered the pain of that torn muscle, afraid that if he took time off from work to have the hernia repaired, he would be fired. After he finally did retire from the Shops in 1952, with forty-five years of service, the hernia was repaired at the Santa Fe Hospital and at the company's expense.[31]

As Albuquerque business exploded, so did its population. "Following World War II, Central Albuquerque experienced an unprecedented population boom. The 1940 census recorded the city's population at 35,449; however, by 1950 the population had more than doubled to 96,815."[32] One of the families counted in that increase was the household of Louis Johnson. He was a native of the Laguna Pueblo village of Paguate, where he had been born in 1920. During the War, he served in the Navy Seabees (Construction Battalions) in the Pacific, and was mustered out of service in the Oakland–Alameda area in California. Because of the already existing Laguna Pueblo satellite village at Richmond, California, associated with the AT&SF locomotive repair shops there, he was able to sign on as a car repairman, under the Watering the Flower Agreement between the tribe and the railroad, discussed earlier. There he married a nurse, and they had their first child.

Around 1949, Louis and his little family returned to New Mexico, where he again got work with the Santa Fe, now as a machinist at the

Albuquerque Shops, where he worked especially on wheels. The Johnsons lived in an upstairs flat on 4th Street in the Barelas neighborhood, above the Wing Ahn Grocery Store. From there, it was just a three-block walk to work for Louis. By now the work schedule at the Shops had settled back to forty hours a week. His oldest daughter remembers each of the children being given a dollar so they could walk to Woolworth's on Central Avenue to have cherry phosphates and grilled cheese sandwiches and then go to a movie. On the weekends during the warm part of the year, and intermittently the rest of the year, the growing family would travel the fifty miles to Laguna Pueblo to help relatives with their farmland.

In 1955, they all moved to a house on Mescalero Avenue in the near Northeast Heights. From then on, until he retired in 1975, Louis commuted to the Shops with three other shopmen. The family now had a car, a turquoise-colored Mercury Comet. Sandra, the daughter, remembers a comfortable life growing up. The children were all sent to parochial school; the family owned their own home, made possible in part because her mother continued to work as a nurse.[33]

Louis Johnson's arrival at the Albuquerque Repair Shops came in the midst of a huge project the Car Repair Shops had been assigned. As reported in February 1948, "The Car Shop has just completed the rebuilding of 1700 obsolete freight cars . . . and is about to begin a new program of 1000 additional cars."[34] That was part of what kept the Shop complex working at a steady pace through the early 1950s and kept optimism high among AT&SF employees and the long list of local businesses that depended indirectly on the Shop payroll. But that was about to change.

CHAPTER 13

The End of Steam

The 1950s

B y 1950, the buzz throughout the railroad industry was about the retirement of steam locomotive fleets and the complete conversion to the use of diesel-electric locomotives. Steam locomotives were rapidly being taken out of service and put into storage or scrapped. Data from the annual reports submitted to the Interstate Commerce Commission (ICC) by the AT&SF reveal the rapid and dramatic shift that occurred between 1949 and 1955. As of January 1, 1949, the Santa Fe owned 1,372 steam locomotives and 644 diesels. But by January 1955, the number of steam locomotives had plummeted to just 134, a decline of 90 percent. Meanwhile, the number of Santa Fe diesels had jumped to 1,618, an increase of 151 percent.[1]

Historian Tom Morrison recently wrote, "The real massacre of the [national] steam locomotive fleet gathered momentum in 1951 to satisfy a national need for scrap metal, partly to build new locomotives and rolling stock, partly to provide munitions for the Korean War."[2] That jibes with our own research on the AT&SF, which began aggressively scrapping steam locomotives in 1949. There were, nevertheless, some Santa Fe steam locomotives repaired at the Albuquerque Shops for use in transporting troops and materiel during the Korean War years of the early 1950s.[3] In any case, the steady decline of numbers of AT&SF steam locomotives meant progressive layoffs of shopworkers at Albuquerque.

The AT&SF was no Johnny-come-lately to dieselization. On the contrary, the Santa Fe had been in the forefront of development of

Table 13.1. AT&SF Number of Steam and Diesel Locomotives, 1949–1955*

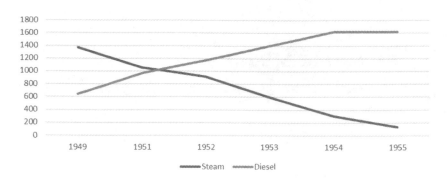

*Data from the annual reports submitted to the Interstate Commerce Commission (ICC) by the AT&SF.

diesel-electric locomotives since the 1920s. As early as 1880, Santa Fe engineers submitted detailed locomotive specifications with their orders to manufacturers. By 1912, however, the company's mechanical engineering department in Topeka was fully responsible for the design of the company's steam locomotives. Those designs, along with "complete sets of specifications and drawings," were then sent to the manufacturers for fabrication.[4] The Santa Fe's longtime chief mechanical engineer Charles Ripley has been especially recognized for his "unsung contributions and achievements in both steam locomotive and Diesel locomotive development [which] were substantial in the heady days leading up to the transition from steam to Diesel locomotives" on the AT&SF.[5]

Design and development of the diesel engine continued into the 1930s. "The first diesel freight locomotive was delivered to the Santa Fe for road tests in February 1938. . . . its performance encouraged the Santa Fe to place the first order for freight diesels by any railroad in the United States." The years of the Second World War made it "clear that diesels had prevailed over steam by every measure of efficiency."[6] With the end of the War, the AT&SF committed itself to a systemwide conversion to diesel power. Over the next decade, the Santa Fe rapidly made the switch, adding nearly two hundred diesel locomotives to its fleet per year for several years. "In August 1957 after eighty-eight eventful years [as a company], all steam operations ended on the Atchison, Topeka & Santa Fe Railway."[7]

Although not perfect machines, diesel-electric locomotives had many virtues that recommended them to railroad businesses over steam locomotives. Diesels used significantly less fuel to haul the same loads; they didn't have to stop for water every hundred miles or so; maintenance and repair costs of diesels amounted to no more than half of what they were for steam engines; and, most important, diesels required a smaller workforce both to run them and to repair them. Because diesel-electric locomotives operated with less mechanical wear and tear than did steam locomotives, their shop visits for repair and overhaul could be scheduled at longer intervals. Railroads could recoup the purchase price of diesels more quickly than the price of their steam-powered cousins.[8] All in all, railroad balance sheets made it clear which type of motive power would make the most money: diesel.

The combination of lengthened intervals between overhauls for diesels and the large-scale standardization of diesel replacement parts meant that railroads needed fewer skilled machinists and other shopmen. It was no longer necessary for machinists to fabricate replacement parts from scratch. The know-how and mastery needed to rebuild a powerful steam locomotive from the ground up was now largely superfluous. As a consequence, between 1949 and 1955 the number of machinists employed by AT&SF fell 18 percent, from 2,222 to 1,818. Santa Fe boilermakers experienced an even more dramatic decline in employment of 59 percent, from 514 in 1949 to 210 in 1955. The skilled labor that had made trains the nation's dominant mode of transportation and haulage for eighty years was in short order not valued as it had been. As has happened so often over time, professions that had seemed steady and secure suddenly were not. Like tailors, general blacksmiths, harness makers, telephone operators, stenographers, check-out clerks, elevator operators, and a host of other once common professions, skilled steam railroad shopmen were threatened with extinction.

The Albuquerque City Directory for 1950 lists 1,322 employees of

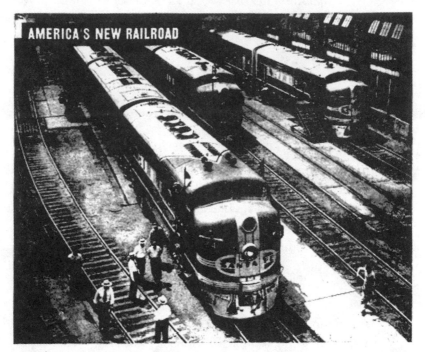

AMERICA'S NEW RAILROAD

Every 3 days a new diesel
joins the Santa Fe fleet !

A 10-year record of "building new" on the Santa Fe

Christened with California champagne, Santa Fe's first multiple-unit diesel locomotive rolled out of Chicago on May 12, 1936.

It powered the first *Super Chief.*

39¾ hours later it rolled into Los Angeles —and the new age in railroading was born.

There were 3600 "horses" in that one.

Five years later, the first multiple-unit freight diesel rolled on Santa Fe rails.

There were 5400 "horses" in that one.

Today, there is more than 2,100,000 *diesel* horse power on the Santa Fe—1324 mighty diesel units.

From 1943 through 1952, a total of 1261 diesel units were placed in service. *Better than one every 3 days for a 10-year record!*

And still they come! 222 in 1953!

Every diesel added, every mile of heavier rail, makes America's New Railroad a little more *completely new.* Why, enough new rail has been laid on the Santa Fe in the last seven years alone to reach *from Chicago to Los Angeles!*

All new—but always the old pride that all this building new costs you, the taxpayer, not one tax penny.

SANTA FE SYSTEM LINES

Albuquerque General Freight and Passenger Offices:
Santa Fe Station, Phone 3 5291

PROGRESS THAT PAYS ITS OWN WAY

Figure 13.1. Newspaper ad for AT&SF's new diesel locomotive fleet, 1953. *Albuquerque Journal,* September 16, 1953, p. 24.

the Locomotive Repair Shops, down modestly from the war years high.[9] But steam locomotives were already then being taken out of service and placed in storage around the country, and a corresponding downward slide in the size of the workforce of shopmen in Albuquerque was not long in becoming apparent. It was in early 1954 when it became unmistakable that the skilled workforce was in for significant downsizing and that the Shops themselves had a very limited future life expectancy. The first official bad news for Shop employees came on January 8, 1954, with a very small note on the front page of the *Albuquerque Journal*. It states simply, "The Santa Fe Railway's conversion to diesel power here has resulted in the layoff of 22 men in the roundhouse and back shops, officials said Thursday."[10] About a month and a half later, C. R. Tucker, the Santa Fe's vice president for operations, was in Albuquerque, along with seven other company officials. Undoubtedly the most eventful news of their visit was that "complete change to Diesel power from steam locomotives would bring a reduction of about 200 men [at the Shops]. But he said the transfer here of roadwork equipment shops would bring 100 men to this department in the building facilities vacated by the locomotive shops." Tucker also tried to ease the coming financial shock to the city of Albuquerque by suggesting that the AT&SF-owned Alvarado Hotel might be enlarged, something that had previously been studied and rejected by the Railway.[11]

Just a week later, on February 28, D. J. Everett, superintendent of the Shops, let it be known that "Albuquerque is the last Santa Fe shop equipped to make major repairs to steam engines. . . . We expect to have all steam power stored within the next few months. . . . Older type equipment is being sold for scrap or other use."[12] Another week, and more bad news: "The Santa Fe Railway shops here will lay off 70 more men today, making a total of 135 here in two weeks."[13] And still the retirement of steam engines continued relentlessly. In January 1955, it was said that "almost all of the Santa Fe's once extensive stable of fine steam locomotives, supreme or not, was retired or 'Laid Up Good.'"[14] By the final months of 1957, the payroll at the Shops was down to between 600 and 700 workers.[15] Because of the continuing decline in steam locomotive numbers and the corresponding shrinkage of the Repair Shops workforce, AT&SF officials felt obliged to try to calm jitters among the remaining employees. A spokesman announced in 1958 that "he does not know of any plans for moving the railroad's shops out of Albuquerque."[16]

One of the Albuquerque shopmen caught up in the flux of the dieselization layoffs during 1952 or 1953 was Thomas Cordova, who we saw earlier had first hired on with the Santa Fe in 1927 as a boilermaker apprentice. Thomas was devastated and out of work for several months after the layoff. He was fortunate enough to eventually find another job as a welder at a Public Service Company of New Mexico (electric utility) power plant construction site. In 1954, he was called back to work at the Santa Fe Shops. But that didn't last; he was laid off from the Shops a second time in 1958. Once again he got a construction job. After a few months, Thomas again rejoined the workforce at the Albuquerque Repair Shops. Now continuing work as a welder, he was an employee at the Shops until he retired in 1976. According to him, the railroad pension he received was very good, and he got medical care at the Santa Fe Hospital in Albuquerque.[17]

Other shopmen confronted by the dieselization layoffs decided to leave Albuquerque in order to pursue technical careers in places where opportunities for machinists and engineers were booming. Such were the cases of brothers-in-law Ernesto Shaw and Jesse Trujillo. Both Ernesto's father, Bonifacio Shaw, and grandfather, George Shaw, had worked for the railroad, and his dad, a machinist, had helped him get a job at the Albuquerque Repair Shops. But when layoffs began on the Santa Fe as a result of the shift to diesel power, Ernesto migrated to Southern California, where he got a job in the burgeoning aerospace industry.[18]

Santa Fe shopmen like Thomas Cordova and Ernesto Shaw were not the only people whose lives were convulsed by the railroad's conversion to diesel power. Some skeptics about wholesale dieselization argued fiercely that abandoning steam power altogether would prove to be a grave misstep, endangering the country. In that camp, some were convinced that the nation would soon exhaust its in-the-ground supply of petroleum fuel.[19] That would, according to the critics, leave the country without rail transport if the complete abandonment of steam power went on in the 1950s as planned.

Other critics, attuned to the recent worldwide catastrophe of war, suggested that to avoid being caught with too few locomotives in the event that World War III were to break out, a considerable fleet of steam locomotives should be stored and kept at the ready. An article from 1950 argues that an emergency reserve fleet of 10,000 to 15,000 steam locomotives should be stationed strategically around the country and that a group of "heavy repair shops," like the one at Albuquerque, should be perpetually on standby to keep those locomotives running, if need be. "Ten modern regional shops, working 24 hours a day, could turn out perhaps 500 Class 2 repairs a month, and in 20 to 30 months all 10 or 15 thousand 'reserve' steam units could be rebuilt."[20]

Only over time did both of these worries subside. In the meantime, railroads such as the Santa Fe did, indeed, mothball steam locomotives at locations along their rail networks so that they could be returned to active service at times of peak demand or in emergencies. The Santa Fe had stored 164 steam locomotives by the end of February 1954, including 16 at the Albuquerque Shops.[21] And, indeed, in late spring 1955, the AT&SF did bring a handful of steam locomotives out of storage at Albuquerque and put them temporarily to work again. "A group of 50-year-old steam locomotives, which were retired last year by the Santa Fe Railway in favor of modern diesel engines, are being taken out of mothballs this week at the Santa Fe shops here [in Albuquerque]. . . . The locomotives, with a new coat of shiny black paint, will be used temporarily between Clovis [NM] and Amarillo [TX]." The contingent of newly overhauled and refurbished steam locomotives comprised less than twenty engines. They were to substitute for diesels that had been sent to California to help handle an especially large potato harvest there.[22]

Barring some catastrophic emergency, though, the system of maintaining a fleet of steam locomotives in reserve was doomed to abandonment from the first. No railroad would long shoulder the increasingly impractical financial burden of keeping hundreds of steam locomotives off the rails, but yet ready to roll at short notice. Eventually, nearly all the steam engines were scrapped, with a few held onto doggedly as emblems of their past pivotal role in the nation's daily life. Within just a few years, knowledge of how to fabricate a locomotive boiler or time the valves on massive steam cylinders or inspect a crosshead was mostly lost or forgotten. A whole suite of professions simply disappeared from the employment repertoire. Neighborhoods like Barelas and South Broadway slipped toward decline. And towns and cities like Albuquerque reeled from the economic blow for several years and then, if they were lucky, gradually shifted to other lines of work.

For Albuquerque, a "dramatic post-war upsurge [in population] ignited an economic revival for light industrial firms, small manufacturers, and wholesale distribution companies, as well as commercial businesses. In addition, the city experienced a major increase in the tourism industry during the 1950s as a result of national prosperity, improved highways . . . and America's continued love affair with the automobile. This economic revitalization came despite the significant loss of railroad jobs as the AT&SF retired its fleet of steam locomotives."[23] Kirtland Air Force Base and Sandia Base (which became Sandia National Laboratories) were already major employers during the War years and continued to grow, as what President Eisenhower labeled "the military-industrial complex" became and then remained a pivotal driver of the national economy. Federal government agencies, such as the Bureau of Land Management, the National Forest Service, the Bureau of Indian Affairs, and the Atomic Energy Commission (which passed through a number of organizational changes ultimately within the Department of Energy), and corresponding state agencies burgeoned. Outside of governmental entities, businesses engaged in commercial oil and gas exploration multiplied, with Albuquerque serving as a financial hub for much of that activity in New Mexico. The University of New Mexico became a substantial and eminent regional institution of higher education. It wasn't many years before the glory days of the Albuquerque Locomotive Repair Shops were all but forgotten.

Although in 1954 the machine shop at the Albuquerque complex, according to the Shops' superintendent, had "already [been] half converted to repair of the diesels," that was not to be included in the Company's final plans.[24] Instead, the AT&SF designated two other already existing shops, at Cleburne, Texas, and San Bernardino, California, as its diesel repair shops. The Albuquerque Shops were relegated to the repair of track maintenance machinery under the designation "Centralized Work Equipment Shops." The workforce still numbered between 600 and 700 at the end of 1957.[25] But in 1977, when William Clarke retired as superintendent of the Albuquerque Centralized Work Equipment Shops and in 1983 when J. B. Hendrix retired as engineer there, the staff numbered only 173.[26] That is about when the Centralized Shops in Albuquerque shut down operations. One man who was working as a machinist when the Centralized Shops closed for good remembered that the company gave just one day's notice. When that day ended, many workers refused to believe that there would be no work the next day. So they showed up at the regular time the next day and found the gate chained and padlocked. That was the end.

"In August 1982, the roundhouse [had been] closed."[27] "The railroad maintained a presence on the property until they closed their doors and the property was sold to a development group in the early 1990s."[28] Then the Santa Fe merged with the Burlington Northern in 1995, to form the BNSF Railway, which is currently the largest freight railroad in North America. After the naming of the Centralized Work Equipment Shops, known to the company as "Shops #3," in the 1950s, the full complex of buildings that had comprised the Albuquerque Locomotive Repair Shops was no longer useful as originally conceived. The electricity-generating powerhouse and its 230-foot-tall smokestack were torn down in 1984, and the roundhouse was demolished in 1987. The cab paint shop, the General Office Building, and other assorted smaller buildings were also taken down in the 1980s. The machine shop, boiler shop, paint shop, and blacksmith shop were all put to use as part of the Centralized Work Equipment Shops. Parts of those buildings, as well as some smaller buildings, also served as storage spaces.[29]

By the 1970s and 1980s, the residential neighborhoods surrounding what had been the Shops complex, particularly Barelas and South

Table 13.2. Ethnicity (Surname Proxy) of AT&SF Employees Living on South 3rd Street and South Broadway in Albuquerque, 1896, 1919, and 1950. Data from City Directories for those Years

	1896	1919	1950
		SOUTH 3RD STREET	
Hispanic	0%	52%	85.4%
Non-Hispanic	100%	48%	14.5%
		SOUTH BROADWAY	
Hispanic	4.3%	31.4%	67.6%
Non-Hispanic	95.6%	68.6%	32.4%

Broadway, were being abandoned, a little more each year. Those neighborhoods, though, during the War and into the 1950s, had been aging but solidly middle-class areas. Patrick Trujillo, who, in the 1950s, spent time on South Arno Street in the South Broadway neighborhood with his grandparents—his grandfather was a machinist at the Shops—remembered the environs as dominantly Hispanic, very tidy, and well maintained, fences neatly painted, yards planted with flowers.[30]

The Hispanization over time of AT&SF employees living in both the Barelas and South Broadway neighborhoods is obvious from our tabulation of families listed by surname in city directories for three different years during operation of the Repair Shops: 1896, 1919, and 1950. The percentage of Spanish-surnamed and non-Spanish-surnamed households on South 3rd Street in Barelas in 1896 was 0 percent Hispanic and 100 percent non-Hispanic, but in 1950 the percentages had almost reversed: 85.4 percent Hispanic and 14.5 percent non-Hispanic. The results are similar for South Broadway, a neighborhood where most people worked at the Shops and made living wages.

After the radical downsizing of the payroll at the Albuquerque Repair Shops in the 1950s, alternative employment and new shopping and residential areas elsewhere in the city—especially on the East Mesa—encouraged a steady outflow of people from the neighborhoods surrounding the rail yards. As we have already seen, there was also significant out-migration from the city as a whole. By the 1960s, "dense neighborhoods north and south of the [central business district, including Barelas and South Broadway] were beginning to look less well kept than in years past" and "shabbiness had slowly given way to dilapidation."[31] There were increasing numbers of derelict and empty houses. Urban renewal efforts of the 1970s, which were never brought to fulfillment, resulted in the tearing down of many houses, leaving empty lots, some of which have yet to be rebuilt upon. A major exception to that trend has been the establishment of the National Hispanic Cultural Center, which opened on a twenty-acre campus within the Barelas neighborhood in 2000.

CHAPTER 14

Shopwomen and African American and Hispanic Shopmen

Throughout the working life of the Albuquerque Locomotive Repair Shops, employees there were known collectively as "shopmen." Despite that designation, women also always made up part of the workforce. Usually, women filled what were traditionally office jobs, such as clerks, stenographers, and secretaries. In those capacities, women were employed in the master mechanic's office, the timekeeper's office, the car accountant's office, and especially in the store-supply depot. Nevertheless, records from 1896 to 1950 show that there regularly were exceptional women who worked side by side with men on the shop floor. All such examples that we have located of women working directly in repairing and overhauling locomotives involved women classified as helpers, commonly machinist helpers. That put women in situations where precision metal working, application of brute force to parts and tools, and routine exposure to physical risk and sexual harassment were all part of the workday.

In 1919, of the four women working on the shop floor, two were single and two were married; three of them were machinist helpers and the fourth was an air brake repair helper. Another seventeen women did office work, eight of them in the storehouse office. Timely receipt of thousands of parts and bulk materials in quantities that met the pressing needs of scheduled maintenance and repairs was always critical to smooth operation of the Shops. Female employees were frequently charged with seeing that the store was fully stocked as a matter of routine.

Although we were not able to locate women to interview who had worked at the Albuquerque Locomotive Repair Shops, we did talk extensively with the daughter of a woman who worked at the AT&SF roundhouse in Winslow, Arizona. Helen Toya, a native of Seama village at Laguna Pueblo—and grandmother of Debra Haaland, US Representative from the First Congressional District in New Mexico—started working at the Winslow yards in the early 1920s, along with her husband Tony, who was from Jemez Pueblo. They had both been signed on by a Santa Fe recruiter during the March feast day at Laguna. At Winslow Helen worked many jobs, from cleaning locomotives to cleaning troop cars during World War II. She was foreman of the car cleaners and often worked the graveyard shift—midnight till 8:00 a.m. At times she also worked at La Posada, the Harvey House hotel in Winslow. For years Helen and Tony lived in the so-called Indian village near Winslow, which consisted of box-car homes. Each home was composed of two box cars joined together at right angles, forming a *T*. Some fifty Native families lived in the Winslow Indian Village. Helen's daughter, Mary Toya, remembers that her family heated their box-car home with a wood stove. They would travel in the direction of the San Francisco Peaks near Flagstaff, Arizona, to cut their fuel wood. Eventually, the Toya family, along with one of Helen's sisters, moved to Desert View Estates on the edge of Winslow. Tony, who also served in several different job classifications, finally retired from the AT&SF after forty-five years, in 1969 or 1970.[1]

Because of the attention that has been paid in recent years to women who worked in manufacturing jobs during World War II (Rosie the Riveters), we gave special scrutiny to the employment information recorded in the *Albuquerque City Directory* for 1943, approximately in the middle of US involvement in the Second World War. Based on a 44 percent sample of the 1943 *City Directory*, we estimate that there were only eleven women working at the shops in 1943, almost all of them in office jobs as clerks and stenographers, while just two worked on the shop floor. One of those women on the floor was a machinist helper named Elisha H. Baker. Indicative of the low ratio of women to men at the Shops during the War is the accompanying photo from a

Figure 14.1. Presentation of the Minute Man flag at Santa Fe Shops, June 1943. The award was to the staff of the store department for exemplary participation in a war bond drive. Note two women employees near the center of the photo. Photographer unknown. Courtesy of the Albuquerque Museum, Photo Archives, Catalog No. PA1980.184.686.

war bond drive among the store department staff in the 1940s, which shows just two women and forty-eight male employees. Surprisingly, we found a lower rate of employment of women in the Albuquerque Locomotive Repair Shops for 1943 (eleven) than for either 1919 (twenty-one) or 1950 (nineteen).

Meanwhile, the city directory listed many women as employed during 1943 in other lines of work in Albuquerque, especially as teachers, nurses, clerks, stenographers, and waitresses, but also in local and federal government jobs, in banks, and as telephone operators and bus drivers. More in line with our original expectations, there were also a few women employed in manufacturing. For example, the *City*

Directory for 1943 lists four women working as welders and one as a sheet metal worker, but none of them were at the Locomotive Repair Shops.[2]

Despite a lack of readily available records of women working in the Albuquerque Shops in 1943, the AT&SF released a promotional film in 1944 that boasted about female shopworkers within the Santa Fe system generally. The narrator of the film, titled *Loaded for War*, proclaimed: "In ever increasing numbers women are helping the railroads to maintain their high standards of service to the nation. But don't think that the only jobs women are performing are behind office desks. With thousands of rail employees going into the armed forces, women readily volunteered for heavier duties in the shops. They are holding down man-sized jobs and handling those jobs with comparable skill."[3]

By comparison, during World War II, railway lines in Great Britain saw a huge influx of women into the workforce, including in service and repair shops. Historian Susan Major has recently shown, for example, that "during the war there had been 39,000 women working for LMS [London, Midland, and Scottish Railway], around 17 percent of the total staff."[4] At the time of the War, LMS was the largest commercial enterprise in the British Empire. The percentage of women railroad workers at LMS was close to the average for all British railroads, as determined in 1944, which was far higher than the comparable rate at the Albuquerque Locomotive Repair Shops. Nevertheless, there had been considerable grousing about female railroad workers by male staff on British railroads throughout the War.[5]

Rather than a verifiable surge in employment of women at the Albuquerque Shops during the War, we found a further increase in the number of Hispanic shopmen, a continuation of the trend that had held since the end of the first decade of the Shops' existence. In 1943, according to the *City Directory*, 63.1 percent of AT&SF shopworkers in the categories of blacksmith, blacksmith apprentice, blacksmith helper, boilermaker, boilermaker apprentice, and boilermaker helper at Albuquerque were Spanish-surnamed. That compares to 28.8 percent Spanish-surnamed in 1919, 50.3 percent in 1925, and 63.1 percent in 1950. Tempering the sense of progress for Hispanos at the Shops through the years is the fact that in 1943 the vast majority of Hispanic shopmen (67 out of 87) were still found among the helpers, the lowest pay grade among machinists and boilermakers (see appendix 2).[6]

Ironically, the percentage of Hispanic shopmen at Albuquerque reached its zenith in the decade or so immediately preceding the end of steam locomotive repair and the collapse of employment for steam locomotive machinists and boilermakers of whatever ethnicity throughout the United States. The near-total lay-off of shopmen at Albuquerque in the mid-1950s was even more drastic and sudden than the piecemeal implosion of general manufacturing employment in the United States was three to four decades later.

Just as we had expected to find increased employment of shopwomen at Albuquerque during the World War II years, we also presumed that there were larger numbers of African Americans working at the Shops during the War. That, too, we did not find evidence for. Among on-the-road crews, positions as Pullman porters, Pullman maids, dining car servers, and baggage handlers had been and continued to be occupied overwhelmingly by African Americans during World War II.[7] But throughout the nation, jobs relating to the repair and overhaul of steam locomotives remained largely closed to African Americans, with only occasional individuals finding places on the shop floor. That was the case at the Albuquerque Shops also throughout the steam era. Even though by the 1940s there was a very small increase in the number of African Americans working on the shop floor, they still made up only a tiny segment of the total workforce.

In 1919, the recorders for the Hudspeth Directory Company entered information about "race," indicating "(c)" for "colored" for those included in the published directory. Leaving aside the question of how accurate or consistent the recorders were, tallying up the entries that included "(c)" provides a rough indication of the percentage of African American residents of Albuquerque who were employed as shopmen by the Santa Fe. Ten persons were identified in this way, all but one of whom served as helpers or laborers. The tenth individual, a man named George Austin, was a turntable operator at the roundhouse. With a total Shop workforce of about 1,195 at the time, this suggests that

Figure 14.2. Group photo of the 1937 Albuquerque machine shop staff. Note the presence of at least 10 African Americans among the 206 workers pictured. An *x* was inked onto the photo above the head of machinist Bonifacio Shaw (fifth row from the bottom and fifth individual from the right). Photographer unknown. Courtesy of Patrick Trujillo, Shaw's grandson.

African American shopmen made up less than 1 percent of the workforce of the Albuquerque Locomotive Repair Shops in 1919.[8]

We have no strictly comparable data for the period 1941–1945, but in seven unstaged photographs taken at the shops in 1943, we are able to distinguish by skin color nine men who appear to be African American out of a total of 188, which yields 4.8 percent African Americans out of that small sample of undetermined representativeness. Perhaps coincidentally, though, a single posed group photo of workers from the machine shop in 1937 reveals 10 out of 206 to be likely African Americans (again based solely on skin tone), giving a remarkably similar percentage of 4.9 percent. At the very least, taken together, the photos from 1937 and 1943 hint at a significant increase in the number of African American shopmen at Albuquerque between 1919 and 1937, but little change resulting from the onset of the World War.

For Albuquerque as a whole, US Census data show that during the 1940s the African American population jumped by 124 percent, a percentage increase that has since been exceeded only during the immediately subsequent decade of the 1950s, when the number of African Americans swelled by 192 percent. Since then, the African American population of Albuquerque has continued to grow, but it has not kept up with the general population growth of the city. Boosted first by air force and army activities, followed by work related to production of the atomic bomb, the population of Albuquerque as a whole exploded during the 1940s, climbing from nearly 36,000 in 1940 to just a shade under 97,000 in 1950.[9]

In sum, among women and minorities, only employment of Hispanic men increased significantly over the seventy-five years of work in steam locomotive repair at Albuquerque. For others, the changes in shop employment were very small and lacked sizeable trends.

CHAPTER 15

Purchase and Redevelopment of the Rail Yards by the City of Albuquerque

emolition of the grand Alvarado Hotel (Harvey House) by the
AT&SF in 1970 was opposed by many citizens who supported
preserving important architectural elements of Albuquerque's
past.[1] An even larger segment of the population was shocked and
immediately lamented the loss of the landmark building once it was
reduced to a graveled parking lot. But it was too late then. As a result,
in the words of Santa Fe architect Barbara Felix, "The teardown of the
Alvarado Hotel in Albuquerque has truly haunted New Mexico" ever
since.[2] The razing of the Alvarado has become an object lesson for historic preservationists around the state and beyond.

So when the Locomotive Repair Shops complex, not far south of the
Alvarado Hotel site, was offered for sale by BNSF in the early 1990s,
many Albuquerqueans were concerned that the tear-down mentality
that had characterized urban renewal efforts in the 1970s and 1980s
would result in obliteration of every vestige of what had been the main
driver of the city's economic activity for many decades. Eventually, that
might have wiped out even the memory of the vibrancy of the Shops
and their immediate surroundings.

Despite several years of delay, those citizens and officials who recognized the ongoing value of the Shops to the community argued their
case in a practical outline with widespread public support. Jessica Carr,
in her 2005 article in *Alibi*, writes, "In November 2000, the Urban
Council of Albuquerque, a nonprofit community redevelopment

Figure 15.1. Photo taken during demolition of the Alvarado Hotel, a Harvey House hotel in Albuquerque, 1970. Photo by Gordon Ferguson. Courtesy of Palace of the Governors, Photo Archives (NMHM/DCA), Negative No. 058706. (Compare figure 9.2.)

project, bought 27 of the railyard's 40 acres, including the historic railyard buildings."[3] This development continued through the decade. As the Urban Land Institute Advisory Services Panel reports, "In 2007, the city of Albuquerque [under Mayor Martin Chávez] and the WHEELS Museum [headed by Leba Freed, Allen Clark, and Joe Craig] formed a partnership to purchase the rail yards from Old Locomotive Shops, LLC. The acquisition was made possible, in part, by grants from the New Mexico State Legislature and the Office of Governor Bill Richardson. The Albuquerque City Council appropriated more than 50 percent of the total cost to purchase the property, and the city became the new owner of the old Santa Fe Railway rail yards on November 28, 2007."[4]

The total purchase price paid by the City of Albuquerque was approximately $8.5 million.

In March 2008, the City Council established a Rail Yards Advisory Board (RYAB) chaired by Councilor Isaac "Ike" Benton. In 2010, the RYAB issued a Request for Proposals for a master developer of the site. Two years later, the City selected Samitaur Constructs of Culver City, California, as the master developer. Although Samitaur produced an ambitious master development plan document in 2014, little further progress was made. As a result, the City terminated their contract in 2018. Within days, the City made the following announcement: "Mayor [Tim] Keller has made a commitment to the redevelopment of the Rail Yards and recognizes the cultural and historic significance of the property to the entire community."[5]

In keeping with the mayor's statement the previous fall, he included the following text in his list of "Top State Legislative Priorities" for the 2019 legislative session:

> The Rail Yards site will be redeveloped to become a vibrant mixed use center of employment, housing, cultural, and educational uses. It will celebrate the historical and cultural legacy of the site and be connected to the community through on-site activity, public access and transportation options. Funding would be used for environmental remediation, expanding market and event space with improvements to the Flue Shop, and improvements to house CNM's [Central New Mexico Community College's] Film School of Excellence, as well as other site improvements. Redeveloping the 27-acre site as a vibrant destination for both residents and tourists, while preserving its historic architecture, will have a significant economic impact on the entire Central New Mexico region. The City is in the process of preparing the Shops complex for redevelopment that preserves its historical and cultural significance.[6]

Accordingly, Albuquerque is pursuing steps to ready the Shops complex for redevelopment.

Figure 15.2. Photo of the west façade of the WHEELS Museum building, 1100 2nd Street SW, Albuquerque, NM, formerly the storehouse of the Albuquerque Locomotive Repair Shops, 2020. Authors' photo.

Further, WHEELS Museum is now located in the Shops' former storehouse at 1100 2nd Street SW on the Rail Yards site. Its displays focus on rail transportation, the Albuquerque Locomotive Repair Shops, the process of overhauling steam locomotives, and the people who worked at the Shops. The collection of vehicles, objects, photographs, books, and documents at Wheels covers all forms of transportation, including antique autos, buggies, bicycles, surreys, milk wagons, a fire truck, and many model displays of cars and trains. A recent donation to the museum is an eighty-five-foot-long private railroad car, the *Silver Iris*, built in 1951. It represents the first of an anticipated group of such cars to be on display at the Rail Yards site.

CHAPTER 16

Conclusion

The Impact of the Shops on Albuquerque and New Mexico

A s long as people had been in the Rio Grande Valley of New Mexico, they had lived with the awareness that the river would, at least every few years, suddenly rise in a flash flood that could dislocate or even destroy life. Repeatedly, people of the Valley had had to bury dead, live with injury, rebuild homes, and re-create fields and irrigation ditches, all as a result of a torrent of water and debris hitting with little forewarning: smashing through homes; ripping out bridges, levees, and dams; burying fields in sand and mud; drowning people, livestock, and wildlife.

The coming of the railroad to New Mexico in 1880 brought a human flood that, with the suddenness of the rampaging river, permanently upended life both in the Valley and far beyond it. There had been hints, analogous to blackening skies and the rumble of thunder, stretching over the previous years, that a powerful force was aimed at the Valley. Beginning even before the laying of the final rails to the vicinity of the small, overwhelmingly Hispanic farming and freighting community of Albuquerque and its neighboring Pueblo villages, the New Mexico & Southern Pacific Railroad (NM&SP, a subsidiary of the Atchison, Topeka & Santa Fe) produced the first in a long succession of seismic transformations in the economic, social, and political lives of the people of the Middle Rio Grande Valley and New Mexico as a whole.

After years of rumors and visits by survey teams and self-proclaimed advance men, the railhead of the NM&SP was poised at Raton Pass in

December 1878. At Albuquerque "new faces appeared at every turn as travelers came and went with greater frequency. Rents and property values were rising and real estate owners were starting costly improvements."[1]

Nevertheless, it was not at all clear then that Albuquerque would become an important point on the railroad line or, indeed, whether that would be a desirable turn of events. In fact, both the town of Bernalillo to the north and the Indian pueblo of Isleta to the south had been eyed by some railroad planners for years as a potential division point. Much of that dreaming and planning, much of the preparatory fieldwork, even months of the initial grading of roadbeds and laying of track remained beyond the notice and attention of farmers of the Middle Rio Grande Valley.

There were some residents of Albuquerque though, predominantly newcomers from the eastern United States, who were anxious to bring a railroad division point to their town. Adding weight to that possibility, during 1879, a major landowner at Bernalillo, negotiating to sell a considerable block of property to the railroad, asked a price far higher than NM&SP was prepared to pay. As a result, the railroad representatives rejected that town's proposal and moved downriver to Albuquerque.

From early 1879 until April 1880, a trio of Albuquerque speculators-developers with access to cash and credit, Franz Huning, William Hazeldine, and Elias Stover, were aggressively urging small, mostly Hispanic, farmers to sell land to them in the general vicinity of Barelas, a cluster of houses and fields southeast of the village of Albuquerque proper. Their hope was to put together a tract of land large enough to accommodate passenger and freight depots and, most important, locomotive and car repair shops, as well as an area where the future shopmen and their families could have homes.

To what extent the developers offered incentives or applied pressure to the farmers is unknown, but it strains credulity to imagine that the farmers all uniformly jumped at the chance to sell. More recent incidents in New Mexico suggest the likelihood of generalized resistance to abandoning traditional agricultural pursuits in favor of touted, but uncertain, future economic advantages. Within a few months, though, many of the Barelas area families did sell. Momentum begot momentum, and the developers assembled what became known as the Original Albuquerque Town Site.

Selling farmland was relatively easy to rationalize, freighted though it was with potential conflict; another thing entirely was accommodating the wave of new people who came to work and live on what was recently agricultural land. That presented challenges of a far greater magnitude. The numbers of railroad workers and adjunct professionals alone shocked most Albuquerque area natives. These associates of the railroad were already prepared and expecting to handle the hundreds of jobs necessary for running a railroad division point and supplying materials to be shipped on down the line as the laying of track continued toward two junctions with the Southern Pacific Railroad, one at Deming, New Mexico, and the other at Needles, California.

They came in dormitory cars and tent encampments, ready to do the work of the railroad as soon as their feet touched the soil of the wide Rio Grande floodplain. They amounted to a swarm that descended in the blink of an eye, putting up temporary buildings, laying out windrows of steel rail, mounds of ties, rafts of barrels of spikes and steel plates, tools of all sorts, banks of coal, tanks of oil, and the makings of laundries, lunchrooms, barber shops, and saloons. The cooks, dishwashers, laundry workers, cobblers, tailors, seamstresses, blacksmiths, apothecaries, preachers, prostitutes, and more materialized like wild mustard plants in spring.

This horde of "railroad people" came on horse- and muleback, in wagons, and on foot, ahead of the actual railroad, to prepare the way: to lay out and erect a temporary depot; to dig pits for the daily dumping of ash and clinkers from locomotives, and other pits to permit inspection and repair of the undercarriages and running gear of locomotives and rail cars, and to dig yet more pits for human waste and garbage; to clear routes to the bank of the Rio Grande so that working livestock could be watered easily every day; to dig wells, install associated pumps, and build elevated water tanks that would resupply locomotive tenders reliably day in and day out beginning in just a matter of days. And that was the barest beginning. Cooking tents had to be set up, and firewood had to be gathered, cut, and stacked. A switching yard and a supply yard had to be designated, cleared, and roughly leveled. A telegraph line had to be strung. It must have seemed to the longtime residents of Albuquerque that "railroad people" were everywhere, altering their familiar world.

To do all this, the lead crew spread out, denuding a wide swath of terrain roundabout in order to cut and gather many hundreds of posts for fences and corrals; diverting the Barelas irrigation ditch so that its flow would be useful to the railroad; hauling rock and sand from nearby arroyos for use in footings and foundations. Even with all this accomplished, more materials and labor were needed, so a recruiter rode repeatedly to Albuquerque village (Old Town) to hire temporary laborers, draft animals, and vehicles. The effect of just weeks of whirlwind labor was to transform utterly and permanently the physical, economic, and social environments of Albuquerque and its environs so that by the time the first train reached the hastily set up depot on April 22, 1880, the largely agricultural community had already been pulled and pushed into the grasp of an industrial world hitherto known only by rumor, travelers stories, and occasional fabulous printed reports. There was resistance, but no turning back.

Albuquerque's new neighbors were human beings, to be sure, but humans of very different sorts from those who were natives by long inheritance of the Middle Rio Grande Valley. They spoke multiple different languages: English, German, Italian, French, Yiddish, Chinese, and others. Many were recent immigrants to the United States and former African slaves. Almost none spoke Spanish, as the citizens of Albuquerque overwhelmingly did. The "railroad people" included Catholics, Protestants, Jews, and those who adhered to other faiths or none at all. Even those who had come from agricultural backgrounds, though, were habituated to an industrial life. Their days were governed by clocks and steam whistles. For them distance had already been compressed by steam locomotion to a fraction of what it had been just days earlier for Albuquerqueans. Henceforward, the pace of life would be faster and more uniform than before. And the measure of value would be hard currency, almost without rival.

These differences and the commanding weight of industry and the wage economy would quickly impel alteration of routines and habits of long standing. It is no wonder that many Albuquerqueans felt invaded and violated. Despite the ingrained ethnic and racial bias of railroad management, they found it necessary to hire a significant number of local Hispanic laborers, for a long while almost entirely laborers. Earthmoving and the hauling of construction materials were readily assigned to local crews, who, besides, worked for minimum wages. Even low pay, though, permitted purchase of manufactured goods that began streaming into the Valley as rails were still being spiked down. The railroad thus stimulated and financed, in significant part, the very commercial trade that it brought.

Although access to modern manufactured goods was generally welcome, it immediately became obvious that disparity in pay—and therefore in material wellbeing and social status—between locals and "railroad people" would be an enduring condition. Initially, there was a justification for distinction in hiring, pay, and advancement between trained and experienced railroad workers and unskilled locals, involving decided advantages for employees with roots in the eastern United States and Europe. Those advantages persisted long after locals had demonstrated or acquired the technical skills necessary for the highest levels of shopwork.[2]

As pointed out previously, the 1922 Shopmen's Strike substantially altered that situation. By laying off a large number of journeyman machinists and boilermakers at the Shops because they were union members and replacing them with nonunion workers, the railroad's management made room for more local Hispanos among the skilled ranks of the Shops' workforce. That in turn increased the average incomes of Hispanic employees of the Albuquerque Shops, which meant greater opportunities for the children and grandchildren of those increasingly middle-class Hispanic shopmen. The remainder of the 1920s and the twenty-year period from about 1935 until 1955 were times of prosperity and stability for many employees at the Albuquerque Locomotive Repair Shops and their families.

That set the stage for a surge of upward financial and social mobility

among those families and their offspring. Increasingly, children of Albuquerque's Hispanic and Native American shopmen completed high
school, went to college, and joined professions. Descendants of AT&SF
employees whom we interviewed for this book, for instance, include two
judges—including a chief justice of the New Mexico Supreme Court—
several lawyers, an engineer, teachers, a sportscaster, a whole family of
dentists, a vineyard owner, and a physical therapist. We don't want to
imply that children of AT&SF machinists and boilermakers invariably
entered white-collar professions. But, as has been shown for other children and grandchildren of solidly middle-class, skilled industrial workers, an extraordinary number of them did, in fact, do just that.

That particular pathway into the professions, though, disappeared
when skilled industrial jobs were eliminated with the drastic reduction
in staff at the Albuquerque Repair Shops in the middle 1950s. That
same downsizing closed the pipeline for family succession into those
skilled industrial positions. Gone were the days when a machinist
father could put in a good word with a supervisor at the Shops and
launch a son or daughter on a skilled career right in the neighborhood.
As the workforce at the Shops shrank, local businesses that depended
on spending by shopmen began to suffer. It wasn't long before retail
businesses in the Barelas and South Broadway neighborhoods were
closing or moving to more advantageous locations.

The laid-off shopmen themselves sought other skilled mechanical
work at Sandia Laboratories and Los Alamos Scientific Laboratory, the
service and repair shops at the Albuquerque Airport, automobile dealerships, the shops of the Albuquerque bus line, the State Highway
Department, and light manufacturing companies that for several
decades had been sprouting up along the railroad tracks (especially
north of the rail yards). Naturally, though, such industrial employers
within Albuquerque were already generally staffed up and could not
absorb anywhere near all of the suddenly out-of-work shopmen.

Some of them were able to find comparable work at other repair
facilities within the AT&SF system or the shops of rival railroads,
although they were also shrinking their shop staffs. Especially attractive to Albuquerque shopmen were AT&SF's diesel locomotive and car

repair shops in San Bernardino, California, and Cleburne, Texas. But many had to seek work in related businesses outside the railroad industry. Two major employers that offered prospects to former shopmen were oilfield service companies in southeast and northwest New Mexico and elsewhere and the aerospace industry, especially in Southern California. One of the consequences of the drastic down-sizing of the Albuquerque Repair Shops in the mid-1950s, therefore, was a significant out-migration of skilled workers from Albuquerque and New Mexico.

Many are the New Mexico families who now have relatives in Southern California precisely because of the diaspora of shopmen in the 1950s. Andrés Vigil, who worked briefly at the Albuquerque Shops, left to open a grocery store in Barelas. Business at the grocery collapsed, though, with the en masse departure from the neighborhood of hundreds of shopmen and their families. Andrés' son Orlando remembered that the shrinking and subsequent closure of the Shops marked the "beginning of the fall of the neighborhood." He also recalled that many families of former shopmen moved to Albuquerque's South Valley, to Las Cruces, New Mexico, and to California.[3]

In summary, the effects of the Albuquerque Locomotive Repair Shops on the growth and character of the city and its residents can appear as generally positive or mostly negative. Which view dominates depends heavily on what precise slice of time is taken into account, as well as which subgroup of the population is one's focus. On the one hand, without the Shops, Albuquerque would almost certainly be a much smaller community than it is today. Its role in industrial activity; in transportation, warehousing, and distribution of consumer goods; and broadly in the economic life of the state would certainly be much less dominant than it now is. On the other hand, stratification along cultural and ethnic lines within the Shop payroll in terms of income and economic opportunity was probably intensified and rendered more rigid by the institutionalized bias that was part of the baggage that came with the railroad. Hispanic, African American, and Native railroad employees consistently earned less than their counterparts with Northern European pedigrees, reinforcing an already prevalent situation in the US money-based economy. That, in turn, exacerbated social and cultural tensions that flared into open hostility from time to time. Also, the transfer of railroad profits to external financiers and stockholders helped to hold the territory and then state in a perpetual state of near impoverishment, even when its natural resources and the labor of its people helped drive national prosperity. For the railroad, New Mexico was typically considered as a place that had to be passed through on the way to somewhere else, east or west. After World War II, that situation was tempered somewhat by the explosion of tourist visitation, facilitated at first by the railroad.

Exploitation of New Mexico's abundant natural resources and work in a variety of government agencies at the local, state, and national levels fueled a generally rising prosperity for Albuquerqueans and for migrants from smaller communities within the state and from Mexico. That migrant stream has been augmented more recently by people from Vietnam and the Middle East, as well as by internal US migration of retirees and those looking for relief from the rigors of northern winters. Most recently, employment in new technologies and the world of entertainment are drawing further population growth. In one way or another, though, all such change and development rests on a base established by the arrival of the railroad in 1880, the proud labor of shopmen and shopwomen at the Albuquerque Locomotive Repair Shops, and the prosperity enjoyed by them and their families.

Appendix 1
Agreement between William Hazledine and Franz Huning

February 18, 1885

Agreement between William Hazledine and Franz Huning amending an 1881 Contract between Hazledine, Huning, and Elias Stover, giving Huning the right to sell properties listed, and assigning him the responsibility of distributing the proceeds, after deducting his costs, in equal shares to the three partners. The properties to be sold include the following:

Union Depot Frontage (plat filed 2/29/1888)
15 1/3 lots   Asking Price: $1,410 ($92/lot)
Highland Addition, South (plat filed 4/8/1887)
77 lots   Asking Price: $4,950 ($64/lot)
Atlantic and Pacific Addition (no filing date)
76 whole and partial lots   Asking Price: $4,950 ($64/lot)
Huning Highland Addition (plat filed 5/12/1887)
159 lots   Asking Price: $4,045 ($25/lot)
Atlantic and Pacific Addition (no filing date)
259 lots   Asking Price: $7,155 ($78/lot)
Store, Gold Ave. and 1st St.   Asking Price: $5,000
Land next to Brewery   Asking Price: $150
Land in Los Griegos Ranchos   Asking Price: $300
Store in Old Town   Asking Price: $ 500
Land in Los Duranes   Asking Price: $1,000
Land in Los Griegos   Asking Price: $ 250

3 Improved lots in A&P Add.   Asking Price: $1,500 ($500/lot)
El Tajo/Diego Padilla Grant
19,788 varas   Asking Price: $4,947
                                     Total Asking Price: $36,157

Appendix 2
Ethnicity of Shopworkers (Surname Proxy),
AT&SF Albuquerque Locomotive Repair Shops

Percentage of Employees in Six Job Classifications, machinist (journeyman, apprentice, helper) and boilermaker (journeyman, apprentice, helper), as Identified in Annual City Directories

Year	Percentage	Number[†]
1919	28.8% Spanish-surnamed	189/656
1925	50.3% Spanish-surnamed	300/597
1943	63.1% Spanish-surnamed	164/260*
1950	63.1% Spanish-surnamed	337/534

* 44 percent sample.
† Spanish-surnamed Workers/Total Workers

Appendix 3
Guide to Steam Locomotive Components

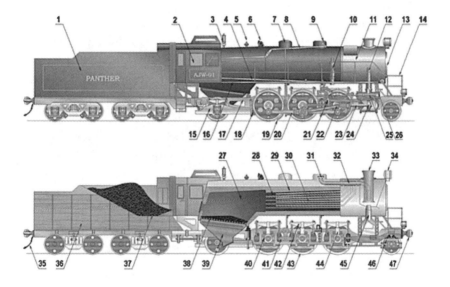

(The image is of a composite imaginary locomotive.)

1. *Tender*. Container holding both water for the boiler and fuel such as wood, coal, or oil for the firebox.
2. *Cab*. Compartment where the engineer and fireman control the engine and tend the firebox.
3. *Whistle*. Steam-powered whistle, located on top of the boiler and used for signaling and warning.
4. *Reach rod*. Rod linking the reversing lever in the cab to the valve gear.

5. *Safety valve*. Pressure relief valve to stop the boiler pressure exceeding the operating limit.
6. *Generator*. Electrical generator driven by small steam turbine for locomotive lighting and headlight.
7. *Sand dome*. Holds sand that is dropped on the rail in front of the driving wheels to improve traction, especially in wet or icy conditions.
8. *Throttle Lever/Regulator*. Sets the opening of the regulator/throttle valve (#31), which controls the pressure of steam entering the cylinders.
9. *Steam dome*. Collects the steam at the top of the boiler so that it can be fed to the engine via main steam pipe, or dry pipe, and the regulator/throttle valve.
10. *Air pump or compressor*. Compresses air for operating the brakes.
11. *Smokebox*. Collects the hot gases that have passed from the firebox and through the boiler tubes.
12. *Steam pipe*. Carries steam to the cylinders.
13. *Smokebox door*. Hinged circular door to allow service access to the smokebox to fix air leaks and remove char.
14. *Handrail*. Support rail for crew when walking along the foot board.
15. *Trailing truck/Rear bogie*. Wheels at the rear of the locomotive to help support the weight of the cab and firebox.
16. *Foot board/Running board*. Walkway along the locomotive to facilitate inspection and maintenance.
17. *Frame*. Carries boiler, cab, and engines and is supported on driving wheels and leading and trailing trucks. The axles run in slots in the frames.
18. *Brake shoe and break block*. Rub on all the driving wheel treads for braking.
19. *Sand pipe*. Deposits sand directly in front of the driving wheels to aid traction.
20. *Side rods/Coupling rods*. Connects the driving wheels together.
21. *Valve gear/motion*. System of rods and linkages synchronizing the valves with the pistons and controls the running and power of the locomotive.

22. *Main rod/Connecting rod*. Steel arm that converts the horizontal motion of the piston into a rotary motion of the driver wheels. The connection between piston and main rod is a crosshead that slides on a horizontal bar behind the cylinder.

23. *Piston rod*. Connects the piston to the crosshead.

24. *Piston*. Driven backward and forward within the cylinder by steam pressure, producing motion from steam expansion.

25. *Valve*. Controls the supply of steam to the cylinders, valve position relative to piston determined by valve gear connected to driving wheel. Steam locomotives may have slide valves, piston valves, or poppet valves.

26. *Valve chest/steam chest*. Valve chamber adjacent to cylinder, contains passageways to distribute steam to the cylinders.

27. *Firebox*. Furnace chamber that is built into the boiler and surrounded by water. Various combustible materials can be used as fuel but the most common were coal, coke, wood, or oil.

28. *Boiler tubes*. Carry hot gases from the firebox through the boiler, heating the surrounding water.

29. *Boiler*. Container almost full of water with air space above. The water is heated by hot gases passing through the tubes, producing steam in the space above the water.

30. *Superheater tubes*. Pass steam back through the boiler to dry and superheat the steam for greater efficiency.

31. *Regulator/Throttle valve*. Controls the amount of steam delivered to the cylinders, one of two ways to vary power of the engine (throttle governing).

32. *Superheater*. Feeds steam back through boiler tubes to superheat (i.e., heat beyond boiling temperature of water at boiler pressure) the steam to increase the engine efficiency and power.

33. *Chimney/Smokestack*. Short chimney on top of the smokebox to carry the exhaust (smoke) away from the engine so that it doesn't obscure the footplate crew's forward view.

34. *Headlight*. Light on front of smokebox to illuminate track ahead and warn approach of locomotive to other track occupants.

35. *Brake hose*. Air or vacuum hose for transmitting brake system pressure/vacuum to train brakes.

36. *Water compartment*. Container for water used by the boiler to produce steam.
37. *Coal bunker*. Fuel supply for the furnace, may be wood, coal/coke, or oil. Fed to the firebox either manually or, for bigger fire grates, by mechanical stoker.
38. *Grate*. Holds the burning fuel and allows ash to drop through.
39. *Ashpan hopper*. Collects the ash from the fire.
40. *Journal box*. Housing for the plain bearing on a driving wheel axle.
41. *Equalizing beams/Equalizing levers/Equalizing bars*. Part of the locomotive suspension system, connected to leaf springs, free to pivot about their center, which is fixed to the frame. Function is to even out weight carried on adjacent axles on uneven or poorly laid tracks.
42. *Leaf springs*. Main suspension springs for the locomotive. Each driver wheel supports its share of the locomotive weight using a leaf spring that connects the axle journal box to the frame.
43. *Driving wheel/Driver*. Wheel driven by the pistons to move the locomotive. Drivers are balanced with weights to reduce unwanted motion of the locomotive. There are three sets of driving wheels in this example.
44. *Pedestal or saddle*. Connects a leaf spring to a driver wheel journal box.
45. *Blast pipe*. Directs exhaust steam up the chimney, creating a draught that draws air through the fire and along the boiler tubes.
46. *Pilot truck/Leading bogie*. Wheels at the front to support weight of boiler front end/smokebox and reduce flanging forces between front driving wheels and rail when rounding curves.
47. *Coupling/Coupler*. Device at the front and rear of the locomotive for connecting locomotives and rail cars together.

Notes

Chapter 1

1. Lucero, "Old Towns Challenged by the Boom Town," 39.
2. Dominguez, *The Missions of New Mexico, 1776,* 151.
3. Hughes, *Doniphan's Expedition,* 111.
4. Magoffin, *Down the Santa Fe Trail and into Mexico,* 152.
5. Boyle, *Los Capitalistas,* 94–96.
6. Gregg, *Commerce of the Prairies,* 1:114.
7. Oliva, *Fort Union and the Frontier Army in the Southwest,* 30.
8. Frazer, *Forts and Supplies,* 1–2.
9. Oliva, *Fort Union and the Frontier Army in the Southwest,* 31, 67, 104.
10. Davis, *El Gringo,* 195.
11. Whipple, "Extracts from the [Preliminary] Report of Explorations for a Railway Route," III:14.
12. US Department of Commerce, Bureau of the Census, "Population of Civil Divisions Less than Counties," *Ninth Census, 1870,* 1:204.
13. Borneman, *Iron Horses,* 10–11.
14. Ibid., 37–41.
15. Reigel, *The Story of the Western Railroads,* 82.
16. Burns, "1870s Railroads, Heading West."
17. Borneman, *Iron Horses,* 63–74.
18. Twitchell, *Leading Facts of New Mexican History,* 2:417.
19. Borneman, *Iron Horses,* 132–33.
20. US Department of Commerce, Bureau of the Census, "Population of Civil Divisions Less than Counties," *Tenth Census, 1880,* 1:263.
21. Westphall, "Albuquerque in the 1870s," 258.
22. Twitchell, *The Leading Facts of New Mexican History,* 2:425–26.
23. Bradley, *The Story of the Santa Fe,* 208.
24. Westphall, "Albuquerque in the 1870s," 262, 264.
25. Ibid., 256.

26. Bryan, *Albuquerque Remembered*, 105; Billington, *New Mexico's Buffalo Soldiers*, 118.

27. Quoted in Westphall, "Albuquerque in the 1870s," 266.

28. Price, *Albuquerque*, 12.

Chapter 2

1. Gregg, *Commerce of the Prairies*, 1:307.

2. Bandelier, *The Southwestern Journals of Adolph F. Bandelier*, 1:69–71.

3. "Steam Locomotive Explosions, 1904," 65.

4. Buell, *Basic Steam Locomotive Maintenance*, 81–85.

5. Wilson, "The Historic Railroad Buildings of Albuquerque."

6. Borneman, *Iron Horses*, 137.

7. Ibid.

8. Ibid., 137–38.

9. US Geological Survey, *Mineral and Water Resources of New Mexico*, 105–12.

10. Ibid., 415, 418.

Chapter 3

1. Westphall, "Albuquerque in the 1870s," 259.

2. Ibid., 261.

3. Ibid., 264.

4. Ibid., 262–63.

5. Bernalillo County, New Mexico, "Huning and Wife to Atlantic & Pacific RR Co.," 25–26.

6. Lucero, "Old Towns Challenged by the Boom Town," 51–52.

7. Westphall, "Albuquerque in the 1870s," 263.

8. Bernalillo County, New Mexico, "Huning to Stover and Hazeldine," 335–38.

9. Bernalillo County, New Mexico, "Phelan and Wife and Huning and Wife to Charles Buford," 274; Bernalillo County, New Mexico, "Huning and Wife to Ferdinand Girard," 183–84.

10. "Inflation Calculator," accessed December 2017, http://www.in2013dollars.com/1880-dollars-in-2015?amount=1.

11. Roberts, comp., *Albuquerque City Directory and Business Guide for 1896*, 133.

12. Bryan, *Albuquerque Remembered*, 106–8.

13. Dodge, "Historic and Architectural Resources," Sec. E:7–8.

14. Ibid., Sec. E:10.

15. "Railway Construction," 634.

16. Twitchell, *Leading Facts of New Mexican History*, 2:426.

17. *Railway Age* 8(4) (January 25, 1883): 53.

18. *Railway Age* 8(11) (March 15, 1883): 144.

19. *Railway Age* 8(28) (July 12, 1883): 407.

20. Bryan, *Albuquerque Remembered*, 107–8, 110.

21. Ibid., 118.

22. Baxter, "Along the Rio Grande," 695.

23. "Santa Fe Completes Modern Shops at Albuquerque," 237; Wilson, "The Historic Railroad Buildings of Albuquerque, An Assessment of Significance, 1986," TMs, p. 1.

24. Crump, Walz, Priest, and Priest, *Santa Fe Locomotive Facilities*, 2:55.

25. Browne, "Notes on Franz Huning," 123.

26. Samitaur Constructs and City of Albuquerque, "Albuquerque Rail Yards Master Development Plan," 8.

27. Ducker, *Men of the Steel Rails*, 27.

28. Simmons, "Trail Dust."

29. Wolberg, "The History of the Streetcar System in Albuquerque, New Mexico," 3.

30. Ibid., 12.

31. Dodge et al., "Atchison, Topeka & Santa Fe Railway Locomotive Shops," Sec. 8:11.

32. KRQE TV News 13, "Historic Albuquerque Trolley Tracks Arrive at Local Museum," May 3, 2017.

33. Roland Johnson, former governor of Laguna Pueblo, interview by Richard and Shirley Flint.

34. Peters, "Watering the Flower," 36.

35. Furze, "The Influence of the Albuquerque Streetcar on the Built Environment," 16.

36. Bryan, *Albuquerque Remembered*, 119.

37. Twitchell, *Leading Facts of New Mexican History*, 2:502, n421.

38. US Department of Commerce, Bureau of the Census, "Abstract," 143.

39. Twitchell, *Leading Facts of New Mexican History*, 2:515, n434.

40. La Farge, *Santa Fe*, 131.

41. Read, *An Illustrated History of New Mexico*, 562–63.

42. US Department of Commerce, Bureau of the Census, "Abstract," 25, 332, 337.

43. "Railway Record," 24.

Chapter 4

1. Keleher, *Memoirs*, 23–25.

2. Read, *An Illustrated History of New Mexico*, 526.

3. Bryan, *Albuquerque Remembered*, 108.

4. Frost, *The Railroad and the Pueblo Indians*, 59.

5. Ibid., 66–67.

6. US Department of Commerce, Bureau of the Census, "Abstract," 32, 55.

7. *Albuquerque Weekly Journal*, December 1, 1882, 3, 4.

8. Twitchell, *Leading Facts of New Mexican History*, 2:491–92.

9. Dodge et al., "Atchison, Topeka & Santa Fe Railway Locomotive Shops," Sec. 8:4–5.

10. Ibid., Sec. 8:3, 11.

11. Ibid., Sec. 8:9.

Chapter 5

1. Dodge et al., "Atchison, Topeka & Santa Fe Railway Locomotive Shops," Sec. 7:6–7.

2. Ibid., 7:7.

3. Ibid.

4. Buell, *Basic Steam Locomotive Maintenance*, 145.

5. Dodge et al., "Atchison, Topeka & Santa Fe Railway Locomotive Shops," Sec. 7:7.

6. Bryant Jr., *History of the Atchison, Topeka and Santa Fe Railway*, 147.

7. Davis, *Power at Odds*, 14.

8. Wright, *1944 Locomotive Cyclopedia of American Practice*, Sec. 20:1, 111–16.

9. Davis, *Power at Odds*, 14.

10. Churella, *From Steam to Diesel*, 10.

11. Davis, *Power at Odds*, 14.

12. Eloy Gutiérrez, interview by Richard and Shirley Flint.

13. James, *Enginemen's Manual*, 414, 415.

14. Buell, *Basic Steam Locomotive Maintenance*, 136; Trains Discussion Forum, July 30, 2011, "Steam Loco Wheel Size," accessed May 6, 2019, https://forums.auran.com/trainz/showthread.php?76884-Steam-Loco-Wheel-Size.

15. James, *Enginemen's Manual*, 414, 415.

16. London, Midland & Scottish (LMS) Railway Shops in Britain, "Locomotive Overhaul," accessed May 6, 2019, http://www.youtube.com/watch?v=7ScBfNGOSiU.

17. Wright, *1944 Locomotive Cyclopedia of American Practice*, Sec. 20: 1,108–16.

18. Davis, *Power at Odds*, 15.

19. "Steam Locomotive Explosions, 1904," 63.

20. Churella, *From Steam to Diesel*, 11.

21. Ducker, *Men of the Steel Rails*, 7.

22. Mike and Leonor Baca, interview by Richard and Shirley Flint.

23. Eloy Gutiérrez, interview by Richard and Shirley Flint.

24. Ducker, *Men of the Steel Rails*, 7.

25. "Among Us," September 1911, 92.

26. "Among Ourselves" (a): 75.

27. "Among Ourselves" (e).

28. Ducker, *Men of the Steel Rails*, 45.

29. Ibid., 43–44.

Chapter 6

1. Churella, *From Steam to Diesel*, 10.

2. Brasher, *Santa Fe Locomotive Development*, Kindle edition image 532.

3. Shipman, "AT&SF Railway, Car Department Apprentice System," 32.

4. "Profitable and Unprofitable Machines," 292.

5. Davids, "Apprenticeship and Guild Control in the Netherlands," 69.

6. Shipman, "AT&SF Railway, Car Department Apprentice System," 32.

7. Ibid., 31.

8. Ibid., 32.

9. Mike and Leonor Baca, interview by Richard and Shirley Flint; Eloy Gutiérrez, interview by Richard and Shirley Flint; Olivia Cordova Loomis, interview by Richard and Shirley Flint.

10. Shipman, "AT&SF Railway, Car Department Apprentice System," 32.

11. Eloy Gutiérrez, interview by Richard and Shirley Flint.

12. Davis, *Power at Odds*, 22–23.

13. Eloy Gutiérrez, interview by Richard and Shirley Flint.

14. Dodge et al., "Atchison, Topeka & Santa Fe Railway Locomotive Shops," Sec. 9:46.

15. *Hudspeth's Albuquerque, Bernalillo County, New Mexico, City Directory*, 1950, passim; also, see especially chapter 10 of this book, "The Railroad Shopmen's Strikes of 1893 and 1922."

16. Roberts, *Albuquerque City Directory and Business Guide for 1896*, passim.

17. Mike and Leonor Baca, interview by Richard and Shirley Flint.

18. Olivia Cordova Loomis, interview by Richard and Shirley Flint.

Chapter 7

1. Westphall, "Albuquerque in the 1870s," 265.

2. Ducker, *Men of the Steel Rails*, 27.

3. Ibid., 25–26, 51.

4. Roberts, *Albuquerque City Directory and Business Guide for 1896*, passim.

5. DeMark, "Occupational Mobility and Persistence Within Albuquerque Ethnic Groups, 1880–1910," 390–91.

6. Ibid., 398.

7. Ducker, *Men of the Steel Rails*, 42–43.

8. While these numbers do not account for all the machinists at the Shops, they do represent a very large sample. It must be noted that throughout this book we employ Spanish surnames as a proxy for Hispanic cultural identity. We recognize that within the group of Spanish-surnamed individuals there were always some Pueblo people who also had Spanish surnames. When looking at Shop employees, though, it is clear that almost always Spanish surnames were associated with Hispanos. There were also Pueblo Shop employees with non-Spanish surnames, exemplified by such men as Louis Johnson, a wheel machinist from Laguna Pueblo, and Lorenzo Jojola, a machinist from Isleta Pueblo, and several other Jojolas from Isleta.

9. Roberts, *Albuquerque City Directory and Business Guide for 1896*, passim; Jens Manuel Krogstad and Mark Hugo López, "For Three States, Share of Hispanic Population Returns to the Past."

10. Ducker, *Men of the Steel Rails*, 28.

11. Judge Joseph Baca, interview by Richard and Shirley Flint; Mike and Leonor Baca, interview by Richard and Shirley Flint.

12. Keleher, *Memoirs*, 11–24; William B. Keleher, interview by Richard and Shirley Flint.

13. Ducker, *Men of the Steel Rails*, 33.

14. Ibid., 49.

15. Ibid., 78.

16. Ibid., 51.

17. Ibid., 45.

18. "Rules and Regulations," 70.

19. Ducker, *Men of the Steel Rails*, 45, 47.

20. Busser, "A Great Moral, Social and Intellectual Movement on the Santa Fe System," 29–32.

21. Ducker, *Men of the Steel Rails*, 53.

22. Ibid., 38–39.

23. Ibid., 10.

24. Frost, *The Railroad and the Pueblo Indians*, 130.

25. AT&SF Railway, *First Annual Report*, 4.

26. Hoffmann, "The Depression of the Nineties," 139, 141, 142.

Chapter 8

1. "Santa Fe Completes Modern Shops at Albuquerque," 239.

2. Eloy Gutiérrez, interview by Richard and Shirley Flint.

3. Olivia Cordova Loomis, interview by Richard and Shirley Flint.

4. Roberts, *Albuquerque City Directory and Business Guide for 1896*, passim; *Worley's Directory, Albuquerque, New Mexico, 1909–10*, passim; *Hudspeth's Albuquerque, Bernalillo County, New Mexico, City Directory, 1950*, passim; *Hudspeth's Albuquerque City Directory, 1919*, passim; *Hudspeth's Albuquerque City Directory, 1925*, passim.

5. Judge Joseph Baca, interview by Richard and Shirley Flint.

6. Patrick Trujillo, interview by Richard and Shirley Flint.

7. Mary Jeannette Swillum Koerschner, phone interview by Richard Flint.

8. Mike and Leonor Baca, interview by Richard and Shirley Flint.

9. Eloy Gutiérrez, interview by Richard and Shirley Flint.

10. Mike and Leonor Baca, interview by Richard and Shirley Flint.

11. Judge Joseph Baca, interview by Richard and Shirley Flint.

12. Joel Tito Ramírez and Carmen Ramírez, recorded interview for the Barelas Oral History Project.

13. Michael Keleher, interview by Richard and Shirley Flint.

14. Judge Joseph Baca, interview by Richard and Shirley Flint.

15. US Congress, *An Act to Establish an Eight-Hour Day for Employees of Carriers Engaged in Interstate and Foreign Commerce*, chapter 436, 721–22; Davis, *Power at Odds*, 37–39.

16. Mike and Leonor Baca, interview by Richard and Shirley Flint.

17. Olivia Cordova Loomis, interview by Richard and Shirley Flint.

18. Ibid.

19. Ducker, *Men of the Steel Rails*, 64.

20. Patrick Trujillo, interview by Richard and Shirley Flint.

21. Judge Joseph Baca, interview by Richard and Shirley Flint.

22. Eloy Gutiérrez, interview by Richard and Shirley Flint.

23. Busser, "A Great Moral, Social and Intellectual Movement on the Santa Fe System," 29.

24. Ducker, *Men of the Steel Rails*, 50.

25. Mary Jeannette Swillum Koerschner, phone interview by Richard Flint.

26. Olivia Cordova Loomis, interview by Richard and Shirley Flint.

27. Sandra Johnson, interview by Richard and Shirley Flint.

28. Mary Jeannette Swillum Koerschner, phone interview by Richard Flint.

29. Mike and Leonor Baca, interview by Richard and Shirley Flint.

30. Tom Turrietta, phone interview by Richard Flint.

Chapter 9

1. Frost, *The Railroad and the Pueblo Indians*, 128.

2. AT&SF Railway, *Twentieth Annual Report*, 13, 21.

3. AT&SF Railway, *Tenth Annual Report*, 28; AT&SF Railway, *Twentieth Annual Report*, 31.

4. AT&SF Railway, *Nineteenth Annual Report*, 21.

5. Ibid.

6. AT&SF Railway, *Tenth Annual Report*, 38; AT&SF Railway, *Twentieth Annual Report*, 41.

7. *Albuquerque Morning Journal.* "Railroads and Shops," Thursday, October 15, 1903, 13.

8. "Illinois Railway Museum Roster: Saint Louis-San Francisco Railroad 1630," accessed May 27, 2019, http://irm.org/cgi-bin/rsearch?steam=St.+Louis-San+Francisco+Railroad=1630.

9. Crump, Priest, and Priest, *Santa Fe Locomotive Facilities*, 1:48, 49.

10. Crump, Walz, Priest, and Priest, *Santa Fe Locomotive Facilities*, 2:55, 56.

11. *Hudspeth's Albuquerque City Directory, 1919*, passim.

12. Mary Jeannette Swillum Koerschner, phone interview by Richard Flint.

13. Patrick Trujillo, interview by Richard and Shirley Flint.

14. US Department of Commerce, Bureau of the Census, "Statistics for New Mexico," 568 and 643; US Department of Commerce, Bureau of the Census. *Fourteenth Census, 1920*, 9:959 and 968.

15. *Hudspeth's Albuquerque City Directory, 1919*, passim.

16. Dodge et al., "Atchison, Topeka & Santa Fe Railway Locomotive Shops," Sec. 8:4–5.

17. Ibid., Sec. 8:5, 14, 15.

18. Wilson, "The Historic Railroad Buildings of Albuquerque, An Assessment of Significance, 1986," TMs, p. 1, image 5.

19. Bryan, *Albuquerque Remembered*, 172.

20. Fergusson, *Albuquerque*, 20.

21. *Hudspeth's Albuquerque City Directory*, 1919, passim.

22. *Albuquerque Morning Journal.* "Railroads and Shops," Thursday, October 15, 1903, 13.

23. AT&SF Railway, *Tenth Annual Report*, 18.

Chapter 10

1. Ducker, *Men of the Steel Rails*, xi.

2. Kern, "A Century of Labor in New Mexico," passim; AT&SF Railway, *Nineteenth Annual Report*, 21.

3. Ducker, *Men of the Steel Rails*, 109.

4. *Las Vegas [NM] Daily Optic*, vol. 14, April 12–25, 1893; *Albuquerque Democrat*, vol. 13, April 18, 1893.

5. Ducker, *Men of the Steel Rails*, 142.

6. Hoffmann, "The Depression of the Nineties," 138.

7. Frost, *The Railroad and the Pueblo Indians*, 134.

8. AT&SF Railway, *First Annual Report*, 4, 24, 35.

9. AT&SF Railway, *Twenty-Sixth Annual Report*, 42.

10. US Congress, *An Act to Establish an Eight-Hour Day for Employees of Carriers*, chapter 436, 721–22; Davis, *Power at Odds*, 37–39.

11. Davis, *Power at Odds*, 36.

12. Ibid., 39.

13. *Proceedings of the Convention, Railway Employees' Department, 1918*, 37.

14. US Congress, *An Act to Provide for the Termination of Federal Control of Railroads and Systems of Transportation*, chapter 91, 456–99.

15. Davis, *Power at Odds*, 52.

16. Ibid., 54.

17. Ibid., 57–60.

18. Ibid., 65; *Albuquerque Morning Journal*, Sunday, July 2, 1922.

19. Davis, "Bitter Conflict," 438.

20. Dodge et al., "Atchison, Topeka & Santa Fe Railway Locomotive Shops," Sec. 9:44.

21. *Albuquerque Morning Journal*, "Worker's Auto in Collision; 'Scab' Yelled," Saturday, July 8, 1922, 3.

22. Frank Archibeque, recorded interview by Frank Saiz for the Barelas Oral History Project.

23. *Albuquerque Morning Journal*, "Guardsmen Are Held Ready to Quell Trouble Due to Strike," Saturday, July 8, 1922, 3.

24. *Albuquerque Morning Journal*, "Machinist Is Attacked," Saturday, July 8, 1922, 3.

25. Jennie Bargas-García, recorded interview by Frank Saiz for the Barelas Oral History Project.

26. Ducker, *Men of the Steel Rails*, 140.

27. *Albuquerque Morning Journal*, "Railway Ready for a Lengthy Strike; Builds Homes for Men," Saturday, July 15, 1922, 1.

28. *Albuquerque Morning Journal*, "Santa Fe Road Is Working 57% of Its Normal Force in Shops," Sunday, July 30, 1922, 1.

29. *Albuquerque Morning Journal*, "Men Wanted . . . Machinists, Boilermakers, Sheet Metal Workers, Electricians, Car Men and Helpers," Sunday, July 23, 1922, 10.

30. *Albuquerque Morning Journal*, "Freight Moving in New Mexico in Big Volume," Sunday, July 23, 1922, 3.

31. *Albuquerque Morning Journal*, "Work Is Begun on Boiler Shop for AT&SF," Sunday, September 24, 1922, 5.

32. Davis, *Power at Odds*, 159.

33. Ibid., 152.

34. *Albuquerque Morning Journal*, "C. B. & Q. Line Signs Contract with Employes [sic]," Saturday, September 16, 1922, 1.

35. Davis, "Bitter Conflict," 453.

36. *Albuquerque Morning Journal*, "Nationwide Injunction Against Rail Strikers Is Issued by U.S. Court," Saturday, September 24, 1922, 1.

37. Davis, *Power at Odds*, 87.

38. Ibid., 159.

39. *Albuquerque Morning Journal*, "Men Gradually Going Back To Work at Shops," Tuesday, August 27, 1921, 1.

40. Davis, *Power at Odds*, 157.

41. Carmen Aragón-Moya, recorded interview by Frank Saiz for the Barelas Oral History Project.

42. Frank Archibeque, recorded interview by Frank Saiz for the Barelas Oral History Project; Frank Archibeque and Rufina Salazar-Montaño, recorded interview by Frank Saiz for the Barelas Oral History Project.

43. *Hudspeth's Albuquerque City Directory*, 1925, passim.

44. Frank Archibeque and Rufina Salazar-Montaño, recorded interview by Frank Saiz for the Barelas Oral History Project.

Chapter 11

1. Brasher, *Santa Fe Locomotive Development*, Kindle edition images 158, 221, 274.

2. Davis, *Power at Odds*, 12.

3. AT&SF Railway, *First Annual Report*; *Tenth Annual Report*; *Twentieth Annual Report*.

4. "Santa Fe Completes Modern Shops at Albuquerque," 237.

5. Wachter, "New Storehouse at Albuquerque," 55–58.

6. Dodge et al., "Atchison, Topeka & Santa Fe Railway Locomotive Shops," Sec. 7:22.

7. Wachter, "New Storehouse at Albuquerque," 55–58.

8. "Among Ourselves" (b): 76.

9. "Construction Notes" (a–e).

10. "Santa Fe Completes Modern Shops at Albuquerque," 237; Dodge et al., "Atchison, Topeka & Santa Fe Railway Locomotive Shops," Sec. 7:19.

11. "Shop and Terminal Construction in 1922," 80.

12. *Albuquerque Morning Journal*, "Work Has Begun on Building of Santa Fe Shops," Tuesday, November 2, 1920, 4.

13. Wilson, "The Historic Railroad Buildings of Albuquerque, An Assessment of Significance, 1986," TMs, p. 1.

14. See, especially, extracts from Emerson's work in "Standardization and Labor Efficiency in Railroad Shops," 783–86.

15. "Santa Fe Completes Modern Shops at Albuquerque," 240.

16. *Albuquerque Morning Journal*, "Men Gradually Going Back to Work at Shops," Tuesday, August 27, 1921, 1.

17. "Santa Fe Completes Modern Shops at Albuquerque," 238–39.

18. Ibid., 238–40.

19. Dodge et al., "Atchison, Topeka & Santa Fe Railway Locomotive Shops," Sec. 7:11.

20. *Albuquerque Morning Journal*, "Santa Fe Will Spend $500,000 on Boiler Shop," Tuesday, May 23, 1922, 3.

21. *Albuquerque Morning Journal*, "Boiler Shop Is Under Way for Santa Fe Here," Wednesday, September 20, 1922, 1.

22. Ibid.

23. Dodge et al., "Atchison, Topeka & Santa Fe Railway Locomotive Shops," Sec. 7: 15, Sec. 9:42.

24. *Hudspeth's Albuquerque City Directory, 1919*, 52.

25. Crump, Priest, and Priest, *Santa Fe Locomotive Facilities*, 1:49, 70, 71.

26. *Albuquerque Morning Journal*, "Thos. La Rue, Aged 42, Slashes Throat with a Razor; Dies Instantly," Wednesday, December 27, 1922, 1.

Chapter 12

1. "Improved Shop Operation at Albuquerque, N.M.," 333–34.

2. Whiter, "Co-operation through Employee Representation," 798.

3. Olivia Cordova Loomis, interview by Richard and Shirley Flint.

4. Stepek, "How America's roaring '20s paved the way for the Great Depression."

5. Dodge et al., "Atchison, Topeka & Santa Fe Railway Locomotive Shops," Sec. 8:15.

6. Ibid., Sec. 8:16.

7. Wolberg, "The History of the Streetcar System in Albuquerque, New Mexico," 12.

8. *Albuquerque Journal*, Ad for Bus Schedule, January 1, 1929, 7; October 6, 1929, 5; February 2, 1931, 5; December 6, 1931, 6.

9. Dodge et al., "Atchison, Topeka & Santa Fe Railway Locomotive Shops," Sec. 8:20–21.

10. Ibid., Sec. 8:21.

11. Mary Jeannette Swillum Koerschner, phone interview by Richard Flint.

12. Dodge et al., "Atchison, Topeka & Santa Fe Railway Locomotive Shops," Sec. 8:18–19.

13. Ibid.

14. "Among Ourselves" (c) (d).

15. "Among Ourselves" (c).

16. Michael Keleher, interview by Richard and Shirley Flint.

17. Dodge et al., "Atchison, Topeka & Santa Fe Railway Locomotive Shops," Sec. 8:23–24.

18. US Department of Commerce, Bureau of the Census, "Value of Output

of Finished Commodities and Construction Materials," 2:699–702; US Department of Commerce, Bureau of the Census, "Railroad Mileage, Equipment, and Passenger Traffic and Revenue: 1890–1970," 729.

19.   US Department of the Interior, National Park Service, *Great Depression Facts*.

20.   "Santa Fe Railroad Shops Are Big Factor in City's Business," n.p.

21.   US Department of Commerce, Bureau of the Census, "Value of Output of Finished Commodities and Construction Materials," 2:699–702.

22.   Waters, *Steel Trails to Santa Fe*, 420.

23.   Marshall, *Santa Fe: The Railroad that Built an Empire*, 302–7.

24.   Waters, *Steel Trails to Santa Fe*, 422–23, 427.

25.   Ibid., 430.

26.   Dodge et al., "Atchison, Topeka & Santa Fe Railway Locomotive Shops," Sec. 9:47.

27.   Ibid., Sec. 9:48.

28.   Eloy Gutiérrez, interview by Richard and Shirley Flint.

29.   Dodge et al., "Atchison, Topeka & Santa Fe Railway Locomotive Shops," Sec. 9:48.

30.   "Growth of Albuquerque is Paralleled by Santa Fe Ry.," n.p.

31.   Mike and Leonor Baca, interview by Richard and Shirley Flint.

32.   Dodge et al., "Atchison, Topeka & Santa Fe Railway Locomotive Shops," Sec. 8:25.

33.   Sandra Johnson, interview by Richard and Shirley Flint.

34.   "Santa Fe Railroad Shops Are Big Factor in City's Business," n.p.

Chapter 13

1.   AT&SF Railway, *Annual Reports of the Atchison, Topeka & Santa Fe Railway Company*, item no. 218470.

2.   Morrison, *The American Steam Locomotive in the Twentieth Century*, 450.

3.   Stagner, "Thirty Years of 4–8–4s," 36.

4.   Brasher, *Santa Fe Locomotive Development*, Kindle edition image 532.

5.   Ibid., image 864.

6.   Borneman, *Iron Horses*, 341.

7.   Brasher, *Santa Fe Locomotive Development*, Kindle edition image 1260.

8.   Churella, *From Steam to Diesel*, 13–14, 16–17, 22.

9.   *Hudspeth's Albuquerque, Bernalillo County, New Mexico, City Directory, 1950*, passim.

10.   *Albuquerque Journal*, "Change to Diesels Lays Off 22 Here," January 8, 1954, 1.

11.   *Albuquerque Journal*, "Study May Result in Alvarado Hotel Being Enlarged," February 21, 1954, 1.

12. *Albuquerque Journal*, "End of the Steam Dynasty on the Santa Fe Is in Sight," February 28, 1954, 26.

13. *Albuquerque Journal*, "Rail Shops Here Slash Force by 70," March 4, 1954, 13.

14. Brasher, *Santa Fe Locomotive Development*, Kindle edition image 788.

15. *Albuquerque Journal*, "Santa Fe Shop Schedules Fund Meet Thursday," October 29, 1957, 17.

16. *Albuquerque Journal*, "Santa Fe Shops to Stay Here," January 24, 1958, 29.

17. Olivia Cordova Loomis, interview by Richard and Shirley Flint.

18. Patrick Trujillo, interview by Richard and Shirley Flint.

19. Waters, *Steel Trails to Santa Fe*, 435.

20. "A Simple Solution to the Motive Power Problem," 43–44.

21. *Albuquerque Journal*, "End of the Steam Dynasty on the Santa Fe Is in Sight," February 28, 1954, 26.

22. *Albuquerque Journal*, "Steam Locomotives Readied for Action," May 18, 1955, 1.

23. Dodge et al., "Atchison, Topeka & Santa Fe Railway Locomotive Shops," Sec. 8:25.

24. *Albuquerque Journal*, "End of the Steam Dynasty on the Santa Fe Is in Sight," February 28, 1954, 26.

25. *Albuquerque Journal*, "Santa Fe Shop Schedules Fund Meet Thursday," October 29, 1957, 17.

26. *Clovis News-Journal*, "Superintendent Retires from the Railroad," November 4, 1977, 9; *Albuquerque Journal*, "Obituary for J. B. Hendrix," August 24, 2014, accessed January 17, 2019, obits.abqjournal.com/obits/print_obit/245120.

27. Crump, Walz, Priest, and Priest, *Santa Fe Locomotive Facilities*, 2:56.

28. Dodge et al., "Atchison, Topeka & Santa Fe Railway Locomotive Shops," Sec. 9:49.

29. Werkema, "Thirty Years Apart at Albuquerque."

30. Patrick Trujillo, interview by Richard and Shirley Flint.

31. Boruff, "Albuquerque Modernism: Downtown Urban Redevelopment, 1960s–1970s."

Chapter 14

1. Mary Toya, interview by Richard and Shirley Flint.

2. *Hudspeth Directory Company's Albuquerque City Directory*, 1943, passim.

3. *Loaded for War*.

4. Major, *Female Railway Workers in World War II*, 29.

5. Ibid., 69–73.

6. The data cited in this chapter derive from *Hudspeth's Albuquerque City*

Directory, 1919; Hudspeth Directory Company's Albuquerque City Directory, 1925; Hudspeth Directory Company's Albuquerque City Directory, 1943; and *Hudspeth's Albuquerque, Bernalillo County, New Mexico City Directory, 1950.*

7. Kornweibel, "The African American Railroad Experience"; Thaggert, "Hand Maidens for Travelers."

8. *Hudspeth's Albuquerque City Directory, 1919,* 129–446.

9. Gibson and Jung, "Historical Census Statistics On Population Totals By Race, 1790 to 1990," and "By Hispanic Origin, 1970 to 1990, for Large Cities and Other Urban Places in the United States."

Chapter 15

1. Simmons, "Trail Dust."

2. Metcalf, "Revitalizing ABQ's 'Spine,'" *Albuquerque Journal,* October 12, 2015.

3. Carr, "Working on the Railroad."

4. "Urban Land Institute Advisory Services Panel," 7.

5. "Mayor Tim Keller to Give Artists Access to the Rail Yards," City of Albuquerque news release, September 21, 2018, accessed July 9, 2019, www.cabq.gov/mayor/news/mayor-tim-keller-to-give-artists-access-to-the-rail-yards.

6. "Tim Keller, Mayor's 2019 Top State Legislative Priorities," accessed July 9, 2019, www.abq.gov/mayor/documents.

Chapter 16

1. Westphall, "Albuquerque in the 1870s," 257.

2. DeMark, "Occupational Mobility and Persistence Within Albuquerque Ethnic Groups, 1880–1910," 398.

3. Orlando Vigil, recorded interview by Frank Saiz for the Barelas Oral History Project.

Bibliography

Published Material

"Among Ourselves" (a). *Santa Fe Magazine* 9, no. 1 (December 1914).

"Among Ourselves" (b). *Santa Fe Magazine* 9, no. 11 (October 1915).

"Among Ourselves" (c). *Santa Fe Magazine* 14, no. 2 (January 1920).

"Among Ourselves" (d). *Santa Fe Magazine* 14, no. 11 (October 1920).

"Among Ourselves" (e). *Santa Fe Magazine* 34, no. 4 (April 1940).

"Among Us." *Santa Fe Employes'* [sic] *Magazine* 5, no. 10 (September 1911).

AT&SF Railway. *Annual Reports of the Atchison, Topeka & Santa Fe Railway Company to the* [Kansas] *State Corporation Commission, 1945–1959.* Kansas State Historical Society, Item No. 218470.

———. *First Annual Report of the Atchison, Topeka & Santa Fe Railway Company for Six Months Ending June 30, 1896.* New York: C. G. Burgoyne, 1896.

———. *Nineteenth Annual Report of the Atchison, Topeka & Santa Fe Railway Company for the Year Ending June 30, 1914.* New York: C. G. Burgoyne, 1914.

———. "Steam Engine Diagrams and Blueprints." Atchison, Topeka and Santa Fe Collection, Railroad, Box 535, Folder 2, Item No. 221763, Kansas State Historical Society.

———. *Tenth Annual Report of the Atchison, Topeka & Santa Fe Railway Company for the Year Ending June 30, 1905.* New York: C. G. Burgoyne, 1905.

———. *Twentieth Annual Report of the Atchison, Topeka & Santa Fe Railway Company for the Year Ending June 30, 1915.* New York: C. G. Burgoyne, 1915.

———. *Twenty-Sixth Annual Report of the Atchison, Topeka & Santa Fe Railway Company for the Fiscal Year Ending December 31, 1920.* New York: C. G. Burgoyne, 1920.

Bandelier, Adolph F. *The Southwestern Journals of Adolph F. Bandelier, 1880–1892.* 4 vols. Edited by Charles H. Lange and Carroll L. Riley. Albuquerque: University of New Mexico Press, 1966–1984.

Baxter, Sylvester. "Along the Rio Grande." *Harper's New Monthly Magazine* 70 (April 1885): 687–700.

Bernalillo County, New Mexico. "Huning and Wife to Atlantic & Pacific RR Co." County Clerk, Deeds, Book M, June 1880.

———. "Huning and Wife to Ferdinand Girard." County Clerk, Deeds, Book O, April 1881.

———. "Huning to Stover and Hazeldine." County Clerk, Deeds, Book M, February 1881.

———. "Plat of Phelan & Huning's Highland Addition in Albuquerque as surveyed by R. L. Davies, C.E., Oct. 16th, 1880." County Clerk, Maps and Plats.

———. "Phelan and Wife and Huning and Wife to Charles Buford." County Clerk, Deeds, Book M, January 1881, 274.

Berger, Tita, and Adam Sullins, comps. *Pedestrians, Streetcars and Courtyard Housing: Past and Future Albuquerques.* Albuquerque: University of New Mexico, School of Architecture and Planning, 2008.

Billington, Monroe Lee. *New Mexico's Buffalo Soldiers, 1866–1900.* Niwot: University Press of Colorado, 1991.

Borneman, Walter R. *Iron Horses: America's Race to Bring the Railroads West.* New York: Little, Brown, 2010.

Boruff, Max. "Albuquerque Modernism: Downtown Urban Redevelopment, 1960s–1970s." City of Albuquerque Planning Department, accessed January 26, 2019, http://albuquerquemodernism.unm.edu/wp/downtown-urban-redevelopment.

Boyle, Susan Calafate. *Los Capitalistas: Hispano Merchants and the Santa Fe Trade.* Albuquerque: University of New Mexico Press, 1997.

Bradley, Glenn Danford. *The Story of the Santa Fe.* Boston: Gorham, 1920.

Brasher, Larry E., with Sam D. Teague. *Santa Fe Locomotive Development: A Pictorial History in Chronological Order, Steam to Diesel 1869–1957.* Amarillo, TX: Signature, 2006; Kindle edition, 2011.

Browne, Lina Fergusson. "Notes on Franz Huning and His Era by His Granddaughter." In Franz Huning, *Trader on the Santa Fe Trail: Memoirs of Franz Huning.* 103–48. Albuquerque: University of Albuquerque, 1973.

Bryan, Howard. *Albuquerque Remembered.* Albuquerque: University of New Mexico Press, 2006.

Bryant Jr., Keith L. *History of the Atchison, Topeka and Santa Fe Railway.* Lincoln: University of Nebraska Press, 1974.

Buell, Dexter C. *Basic Steam Locomotive Maintenance.* Omaha, NE: Simmons-Boardman Books, 1980.

Burns, Adam. "1870s Railroads, Heading West." Accessed May 14, 2019, https://www.american-rails.com/1870s.html.

Busser, S. E. "A Great Moral, Social and Intellectual Movement on the Santa Fe System." *International Railway Journal* 9, no. 6 (March 1902): 29–32.

Carr, Jessica Cassyle. "Working on the Railroad." *Alibi* 14, no. 46 (November 17–23, 2005), https://alibi.com/news/13401/Working-on-the-Railroad.html.

Churella, Albert J. *From Steam to Diesel: Managerial Customs and Organizational Capabilities in the Twentieth-Century American Locomotive Industry*. Princeton, NJ: Princeton University Press, 1998.

"Construction Notes" (a). *Santa Fe Magazine* 9, no. 1 (December 1914).

"Construction Notes" (b). *Santa Fe Magazine* 9, no. 2 (January 1915).

"Construction Notes" (c). *Santa Fe Magazine* 9, no. 5 (April 1915).

"Construction Notes" (d). *Santa Fe Magazine* 9, no. 9 (August 1915).

"Construction Notes" (e). *Santa Fe Magazine* 9, no. 10 (September 1915).

Crump, Russell L., Robert Walz, Stephen Priest, and Cinthia Priest. *Santa Fe Locomotive Facilities*. Vol. 2, *West End Western Lines*. Parkville, MO: Paired Rail Railroad, 2014.

Crump, Russell L., Stephen Priest, and Cinthia Priest. *Santa Fe Locomotive Facilities*. Vol. 1, *The Gulf Lines*. Kansas City, MO: Paired Rail Railroad, 2003.

Davids, Karel. "Apprenticeship and Guild Control in the Netherlands, c. 1450–1800." In *Learning on the Shop Floor: Historical Perspectives on Apprenticeship*. Edited by Bert De Munck, Steven L. Kaplan, and Hugo Soly, 65–84. New York: Berghahn Books, 2007.

Davis, Colin J. "Bitter Conflict: The 1922 Railroad Shopmen's Strike." *Labor History* 33, no. 4 (Fall 1992): 433–55.

———. *Power at Odds: The 1922 National Railroad Shopmen's Strike*. Urbana: University of Illinois Press, 1997.

Davis, W. W. H. *El Gringo, or, New Mexico and Her People*. Santa Fe: Rydal, 1938. Reprint, Chicago: Rio Grande, 1962. Originally published in 1857.

DeMark, Judith L. "Occupational Mobility and Persistence Within Albuquerque Ethnic Groups, 1880–1910: A Statistical Analysis." *New Mexico Historical Review* 68, no. 4 (October 1993): 389–98.

Dodge, William A. "Historic and Architectural Resources of Central Albuquerque, 1880–1970." Sec. E.: Statement of Historic Contexts. National Park Service, National Register of Historic Places Multiple Property Documentation Form, Statement of Historic Contexts, 2012.

Dodge, William A., consultant, Maryellen Hennessy, Edgar Bolesn, and Petra Morris, City of Albuquerque Planning Department, and Steven Moffson, New Mexico Historic Preservation Division. "Atchison, Topeka & Santa

Fe Railway Locomotive Shops," Sections 7–9. National Park Service, National Register of Historic Places Registration Form, 2014.

Domínguez, Francisco Atanasio. *The Missions of New Mexico, 1776.* Edited and translated by Eleanor B. Adams and Fray Angélico Chávez. Albuquerque: University of New Mexico Press, 1975. Originally published 1956.

Ducker, James H. *Men of the Steel Rails: Workers on the Atchison, Topeka & Santa Fe Railroad, 1869–1900.* Lincoln: University of Nebraska Press, 1983.

Emerson, Harrington. "Standardization and Labor Efficiency in Railroad Shops." *Engineering Magazine* 33 (1907): 783–86.

Fergusson, Erna. *Albuquerque.* Albuquerque: Merle Armitage, 1947.

Frazer, Robert W. *Forts and Supplies: The Role of the Army in the Economy of the Southwest, 1846–1861.* Albuquerque: University of New Mexico Press, 1983.

Frost, Richard H. *The Railroad and the Pueblo Indians: The Impact of the Atchison, Topeka and Santa Fe on the Pueblos of the Rio Grande, 1880–1930.* Salt Lake City: University of Utah Press, 2016.

Furze, Michael. "The Influence of the Albuquerque Streetcar on the Built Environment." In *Pedestrians, Streetcars and Courtyard Housing: Past and Future Albuquerques.* Compiled by Tita Berger and Adam Sullins, 15–21. Albuquerque: University of New Mexico, School of Architecture and Planning, 2008.

Gibson, Campbell, and Kay Jung. "Historical Census Statistics on Population Totals by Race, 1790 to 1990," and "By Hispanic Origin, 1970 to 1990, for Large Cities and Other Urban Places in the United States." Population Division, Working Paper No. 76, Table 32: New Mexico–Race and Hispanic Origin. Washington, DC: US Census Bureau, 2005.

"Grading for the Great New Shops in Albuquerque." *Santa Fe Magazine* 9, no. 11 (October 1915): 44; "Group photo of attendees at the first meeting of the storekeepers of the entire Santa Fe System, September 27–29, 1915," 40; 76.

Gregg, Josiah. *Commerce of the Prairies.* 4th ed. 2 vols. Philadelphia: J. W. Moore, 1849.

"Growth of Albuquerque Is Paralleled by Santa Fe Ry." *Albuquerque Progress* 10, no. 5 (June 1943): n.p.

Hoffmann, Charles. "The Depression of the Nineties." *Journal of Economic History* 16, no. 2 (June 1956): 137–64.

Hudspeth Directory Company's Albuquerque City Directory, 1943. El Paso, TX: Hudspeth Directory, 1943.

Hudspeth's Albuquerque, Bernalillo County, New Mexico, City Directory, 1950. El Paso, TX: Hudspeth Directory, 1950.

Hudspeth's Albuquerque City Directory, 1919. El Paso, TX: Hudspeth Directory, 1919.

Hudspeth's Albuquerque City Directory, 1925. El Paso, TX: Hudspeth Directory, 1925.

Hughes, John T. *Doniphan's Expedition*. Cincinnati: J. A. & U. P. James, 1850.

"Improved Shop Operation at Albuquerque, N.M." *Railway Mechanical* 98, no. 4 (1924): 333–34.

James, W. P. *Enginemen's Manual*. 12th ed. Louisville, KY: W. P. James, 1917.

Keleher, William A. *Memoirs: Episodes in New Mexico History, 1892–1969*. Santa Fe: Sunstone, 2008. Facsimile of the 1969 edition.

Kern, Robert. "A Century of Labor in New Mexico." In *Labor in New Mexico: Unions, Strikes, and Social History Since 1881*. Edited by Robert Kern, 3–24. Albuquerque: University of New Mexico Press, 1983.

Kornweibel, Theodore. "The African American Railroad Experience." Interview by Maureen Cavanaugh, KPBS (San Diego), March 23, 2010. Transcript accessed July 2, 2019, www.kpbs.org/news/2010/mar/23/african-american-railroad-experience.

Krogstad, Jens Manuel, and Mark Hugo López. "For Three States, Share of Hispanic Population Returns to the Past." Pew Research Center. Washington, DC, June 20, 2014. Accessed July 8, 2018, http://www.pewresearch.org/fact-tank/2014/06/10/for-three-states-share-of-hispanic-population-returns-to-the-past/.

KRQE TV News 13. "Historic Albuquerque Trolley Tracks Arrive at Local Museum." May 3, 2017.

La Farge, Oliver. *Santa Fe: The Autobiography of a Southwestern Town*. Norman: University of Oklahoma Press, 1959.

Lister, Francis E., comp. *Car Builders Dictionary*. New York: Railway Age, 1909.

Loaded for War. Chicago: Atchison, Topeka & Santa Fe Railway, 1944. Promotional short film.

Lucero, Brian Luna. "Old Towns Challenged by the Boom Town: The Villages of the Middle Rio Grande Valley and the Albuquerque Tricentennial." *New Mexico Historical Review* 82, no. 1 (Winter 2007): 37–69.

Magoffin, Susan Shelby. *Down the Santa Fe Trail and into Mexico: The Diary of Susan Shelby Magoffin, 1846–1847*. Edited by Stella M. Drumm. Lincoln: University of Nebraska Press, 1982.

Major, Susan. *Female Railway Workers in World War II*. Barnsley, South Yorkshire, England: Pen & Sword Books, 2018.

Marshall, James. *Santa Fe: The Railroad that Built an Empire*. New York: Random House, 1945.

Morrison, Tom. *The American Steam Locomotive in the Twentieth Century*. Jefferson, NC: McFarland, 2018.

Oliva, Leo E. *Fort Union and the Frontier Army in the Southwest*. Southwest Cultural Resources Center, Professional Papers No. 41. Santa Fe, NM: National Park Service, Division of History, 1993.

Peters, Kurt Michael. "Watering the Flower: The Laguna Pueblo and the Atchison, Topeka and Santa Fe Railroad, 1880–1980." PhD diss., University of California, Berkeley, 1994.

Price, V. B. *Albuquerque: A City at the End of the World*. 2nd ed. Albuquerque: University of New Mexico Press, 2003.

Proceedings of the Convention, Railway Employees' Department, 1918. St. Louis, MO, April 8, 1918.

"Profitable and Unprofitable Machines." *Railway Age Monthly and Railway Service Magazine* 1, no. 5 (May 1880): 292.

Railway Age 8, no. 4 (January 25, 1883).

Railway Age 8, no. 11 (March 15, 1883).

Railway Age 8, no. 28 (July 12, 1883).

Railway Age 73, no. 6 (August 5, 1922): 238–39.

"Railway Construction." *Railway Age Monthly and Railway Service* 1, no. 10 (October 1880): 633–37.

"Railway Record." *Railway Age* 8, no. 2 (January 11, 1883): n.p.

Read, Benjamin M. *An Illustrated History of New Mexico*. Santa Fe: New Mexican, 1912.

Reigel, Robert Edgar. *The Story of the Western Railroads*. New York: MacMillan, 1926.

Roberts, C. O'Connor, comp. *Albuquerque City Directory and Business Guide for 1896*. Albuquerque: Hughes & McCreight, 1896.

"Rules and Regulations of the A.T.&S.F. Hospital Association." *Railway Surgeon* 1, no. 3 (1894): 70.

Samitaur Constructs and City of Albuquerque. "Albuquerque Rail Yards Master Development Plan." Albuquerque: 2014.

"Santa Fe Completes Modern Shops at Albuquerque." *Railway Age* 73, no. 6 (August 5, 1922): 237–42.

"Santa Fe Railroad Shops Are Big Factor in City's Business." *Albuquerque Progress* 15, no. 2 (February 1948): n.p.

Shipman, H. L. "AT&SF Railway, Car Department Apprentice System." *Railway Mechanical Engineer* 97, no. 1 (January 1923): 31–33.

"Shop and Terminal Construction in 1922." *Railway Mechanical Engineer* 97, no. 2 (1922): 80.

Simmons, Marc. "Trail Dust: Once Grand Alvarado Hotel Now a Fading Memory." *Santa Fe New Mexican*, September 26, 2014.

"A Simple Solution to the Motive Power Problem." *Railway Age* 128, no. 15 (April 15, 1950): 43–44.

Stagner, Lloyd. "Thirty Years of 4-8-4s." *Trains* (February 1987): 24–36.

"Steam Locomotive Explosions, 1904." *The Locomotive*. 25, no. 2 (April 1904): 63–65.

Stepek, John. "How America's Roaring '20s Paved the Way for the Great Depression." *Money Week*, November 10, 2017, accessed December 6, 2018, https://moneyweek.com/476326/.

Thaggert, Miriam. "Hand Maidens for Travelers: The Missing Story of the Pullman Maids." Lecture presented March 10, 2016, Ruggles Hall, Newberry Library, Chicago, IL.

Twitchell, Ralph Emerson. *Leading Facts of New Mexican History*. 5 vols. Cedar Rapids, IA: Torch, 1911, 1912. Reprint, 2 vols. Albuquerque: Horn and Wallace, 1963.

US Congress. *An Act to Establish an Eight-Hour Day for Employees of Carriers Engaged in Interstate and Foreign Commerce, and for Other Purposes, September 3, 1916*. 64th Cong., 1st sess., H.R. 17700. Public Law 252, chapter 436, 721–22.

US Congress. *An Act to Provide for the Termination of Federal Control of Railroads and Systems of Transportation; to Provide for the Settlement of Disputes Between Carriers and Their Employees; to Further Amend an Act Entitled "An Act to Regulate Commerce," Approved February 4, 1887, as Amended, and for Other Purposes*. 66th Cong., 2nd sess., H.R. 10453, Public Law 66–152, chapter 91, 456–99.

US Department of Commerce, Bureau of the Census. "Abstract." In *Twelfth Census, 1900*. Washington, DC: Government Printing Office, 1902.

US Department of Commerce, Bureau of the Census. *Fourteenth Census of the United States Taken in the Year 1920*. Vol. 9: *Manufactures, 959, 968*. Washington, DC: Government Printing Office, 1923.

US Department of Commerce, Bureau of the Census. "Median Money Income of Families and Unrelated Individuals." In Series G, *Historical Statistics of the United States, Colonial Times to 1970*, bicentennial ed., 2 vols., part 1:297, column 199. Washington, DC: Bureau of the Census, 1975.

US Department of Commerce, Bureau of the Census. "Population of Civil Divisions Less than Counties." Table 3, 1:204. In *Ninth Census, 1870*. 2 vols. Washington, DC: US Government Printing Office, 1872.

US Department of Commerce, Bureau of the Census. "Population of Civil Divisions Less than Counties." Table 3, 1:263. In *Tenth Census, 1880*. 2 vols. Washington, DC: US Government Printing Office, 1882.

US Department of Commerce, Bureau of the Census. "Railroad Mileage, Equipment, and Passenger Traffic and Revenue: 1890–1970." Series Q 284–312, *Historical Statistics of the United States, Colonial Times to 1970*, bicentennial ed., part 2:729. Washington, DC: Bureau of the Census, 1975.

US Department of Commerce, Bureau of the Census. "Statistics for New

Mexico." In *Thirteenth Census of the United States Taken in the Year 1910*. Washington, DC: Government Printing Office, 1913.

US Department of Commerce, Bureau of the Census. "Value of Output of Finished Commodities and Construction Materials Destined for Domestic Consumption at Current Producers' Prices, and Implicit Price Indexes for Major Commodity Groups (Shaw): 1869 to 1939." In Series P 318–74, *Historical Statistics of the United States, Colonial Times to 1970*, bicentennial ed., 2 vols., part 2:699–702. Washington, DC: Bureau of the Census, 1975.

US Department of the Interior, National Park Service. *Great Depression Facts*. Poughkeepsie, NY: Franklin D. Roosevelt Presidential Library and Museum, accessed December 12, 2018, https://fdrlibrary.org/great-depression-facts.

US Geological Survey. *Mineral and Water Resources of New Mexico*. 3rd printing. Albuquerque: New Mexico State Bureau of Mines and Mineral Resources, 1972.

"Urban Land Institute Advisory Services Panel." *Albuquerque Rail Yards, Albuquerque, New Mexico: Redeveloping the City's Historic Rail Yards*. Washington, DC: Urban Land Institute, 2008.

Wachter, A. B. "New Storehouse at Albuquerque, a Model of Efficiency." *Santa Fe Magazine* 9, no. 12 (November 1915): 55–58.

Waters, L. L. *Steel Trails to Santa Fe*. Lawrence: University of Kansas Press, 1950.

The Way Things Work: An Illustrated Encyclopedia of Technology. 2 vols. New York: Simon and Schuster, 1967.

Werkema, Evan. "Thirty Years Apart at Albuquerque." August 8, 2009, accessed January 23, 2019, https://www.trainorders.com/discussion/read.php?11,2000383.

Westphall, Victor. "Albuquerque in the 1870s." *New Mexico Historical Review* 23, no. 4 (October 1948): 253–68.

Whipple, A. W. "Extracts from the [preliminary] Report of Explorations for a Railway Route, Near the Thirty-Fifth Parallel of North Latitude, from the Mississippi River to the Pacific Ocean." In *Reports of Explorations and Surveys to Ascertain the Most Practicable and Economical Route for a Railroad from the Mississippi River to the Pacific Ocean*, 33d Cong., 2nd sess., Senate Ex. Doc. No. 78, III:6–14. 6 vols. Washington, DC: Beverley Tucker, 1856.

Whiter, E. T. "Co-operation Through Employee Representation." *Railway Mechanical Engineer* 97, no. 12 (1923): 798–99.

Wilson, Chris. "The Historic Railroad Buildings of Albuquerque: An Assessment of Significance, 1986." Accessed September 2016, http://wheelsmuseum.org/wp-content/uploads/2015/11/The-Historic-Railroad-Buildings-of-Albuquerque.pdf.

Wolberg, Hannah. "The History of the Streetcar System in Albuquerque, New Mexico." In *Pedestrians, Streetcars and Courtyard Housing: Past and Future Albuquerques*. Compiled by Tita Berger and Adam Sullins, 3–14. Albuquerque: University of New Mexico, School of Architecture and Planning, 2008.

Worley's Directory, Albuquerque, New Mexico, 1909–10. Dallas: John F. Worley Directory, 1909.

Wright, Roy V., ed. *1944 Locomotive Cyclopedia of American Practice*. 12th ed. New York: Simmons-Boardman, 1944.

Interviews

Aragón-Moya, Carmen. Recorded interview by Frank Saiz for the Barelas Oral History Project, December 21, 2001, National Hispanic Cultural Center, Albuquerque, NM.

Archibeque, Frank. Recorded interview by Frank Saiz for the Barelas Oral History Project, December 7, 2001, National Hispanic Cultural Center, Albuquerque, NM.

Archibeque, Frank, and Rufina Salazar-Montaño. Recorded interview by Frank Saiz for the Barelas Oral History Project, December 9, 2001, National Hispanic Cultural Center, Albuquerque, NM.

Baca, Judge Joseph. Interview by Richard and Shirley Flint, July 12, 2017, Albuquerque, NM.

Baca, Mike and Leonor. Interview by Richard and Shirley Flint, August 9, 2017, Los Chávez, NM.

Bargas-García, Jennie. Recorded interview by Frank Saiz for the Barelas Oral History Project, November 27, 2001, National Hispanic Cultural Center, Albuquerque, NM.

Gutiérrez, Eloy. Interview by Richard and Shirley Flint, July 8 and August 16, 2017, Albuquerque, NM.

Johnson, Roland, former governor of Laguna Pueblo. Interview by Richard and Shirley Flint, April 10, 2017, Albuquerque, NM.

Johnson, Sandra. Interview by Richard and Shirley Flint, March 19, 2017, Laguna Pueblo, NM.

Keleher, Michael. Interview by Richard and Shirley Flint, May 25, 2017, Albuquerque, NM.

Keleher, William B. Interview by Richard and Shirley Flint, May 25 and June 22, 2017, Albuquerque, NM.

Koerschner, Mary Jeannette Swillum. Phone interview by Richard Flint, June 26, 2017, Houston, TX.

Loomis, Olivia Cordova. Interview by Richard and Shirley Flint, October 11, 2017, Albuquerque, NM.

Ramírez, Joel Tito, and Carmen Ramírez. Recorded interview for the Barelas
    Oral History Project, December 16, 2001, National Hispanic Cultural
    Center, Albuquerque, NM.
Toya, Mary. Interview by Richard and Shirley Flint, June 7, 2017, Mesita, NM.
Trujillo, Patrick. Interview by Richard and Shirley Flint, September 14, 2017,
    National Hispanic Cultural Center, Albuquerque, NM.
Turrietta, Tom. Phone interview by Richard Flint, September 7, 2017, Albuquer-
    que, NM.
Vigil, Orlando. Recorded interview by Frank Saiz for the Barelas Oral History
    Project, December 4, 2001, National Hispanic Cultural Center, Albuquer-
    que, NM.

Index